ADVANCED
TECHNOLOGY CONCEPTS
FOR COMMAND AND CONTROL

Advanced Technology Concepts for Command and Control

ISBN :   Hardcover   1-4134-1734-5
                Softcover   1-4134-1733-7

This book was printed in the United States of America.

To order additional copies of this book, contact:
Xlibris Corporation
1-888-795-4274
www.Xlibris.com
Orders@Xlibris.com
20411

CONTENTS

Part II: Military Intelligence

Part IV: Operational Command and Control

Part V: Tactical Command and Control

Part VI: Conclusions

ACKNOWLEDGEMENTS

This book is a result of work performed under the Joint Force Air Component Commander (JFACC) Program conducted by the Defense Advanced Research Projects Agency (DARPA). As a part of the large program, this was a collective effort in more than one respect. Not only was the authorship of the book collective, as explained in *About the Authors*, but there were also many other individuals without whose contributions this work would not have succeeded.

Major Sharon Heise of DARPA/ITO, as well as Mike Ownby and Jackie Lynch of Emergent Information Technologies, Inc., and Dr. Steve Morse of SPARTA, Inc. provided support, guidance, and endless hours of reading, rereading, and commenting on the information contained within.

Dr. Tim Busch and Carl Defranco, both of Rome Air Force Research Lab, and Brian Canino and Julie Howell, both of SPAWAR Systems Center, read and commented on both the technical accuracy and military relevance of the book's content.

Dr. Abby Mendelson and Mary Ann Pike made incalculable contributions, massaging, molding, and tweaking a text written by many authors into a coherent whole.

Dulce M. Balicki and Valarie Tassari, both of BBN Technologies, Inc. performed both document preparation and behind-the-scenes production. Although they are nowhere credited as authors, their names are implicit on every page.

Finally, and most important, the researchers, from the various corporations and academic institutions who took part in the JFACC Program, both developed and explained to the authors the technologies presented in these chapters. While some of these extraordinary people are individually recognized in *About the Authors,* space did not permit mentioning all the researchers associated with the development of the technologies mentioned in this book.

PREFACE

The origin of this book lies in a program sponsored by the Defense Advanced Research Projects Agency (DARPA) to explore innovative technologies with the potential to catalyze a revolutionary change in command and control (C^2), specifically joint aerospace operations: the Joint Force Air Component Commander (JFACC) program.

The JFACC program was initiated in 1996, under an original charter of developing technologies for assisting in the rapid, semi-automated construction of a fully delineated, cross-functional air campaign plan. After less than two years of using then-current AI techniques in planning and scheduling, the program decided to abandon this more traditional approach, and in late 1999 adopted a new technical direction. Led by Program Manager Colonel Dan McCorry, USAF, the program recreated itself under the following hypothesis: *Control Theory and its supporting technologies have developed sufficiently to be applicable to problems in military C^2*. It is upon the results of this latter phase of the JFACC program that this book is based.

In a military context, *control* has to do with the commander's assurance that the forces under his or her supervision are properly synchronized in space, time, and purpose to achieve the desired intent. In the context of the engineering discipline of control science, however, the meaning of *control* refers to the principles and methods used to design systems that maintain high performance by automatically adjusting to changes in the environment. Traditional applications for control theory include aircraft flight control systems, control of chemical processes in refining, and control of robotics in manufacturing. The JFACC program conjectured that these mathematically, conventionally, physically based approaches could be applied to the highly uncertain, complex, and often chaotic actions in a military battlespace. The question remaining to be answered, and that we hope to address here, is, if such application provides significant military value.

The JFACC program cannot claim to be the sole pioneer with respect to the application of control technology to C^2. In fact, there exists a substantial body of research developed over the past 25 years to which this work is related, and upon which it partially depends. A watershed event in research on C^2 was a series of workshops, beginning in 1978, sponsored by the Office of Naval Research and the Electronic Systems Laboratory at MIT to address command, control, and communications (C^3) systems [1]. A further workshop in 1979, sponsored by the National Defense University on *Quantitative Assessment of Utility of Command and Control Systems* is credited with "establishing the starting place and time" on the quest for a theory of C^3 [2]. Together with various meetings, these workshops were responding to an accelerating need for coordination and control of C^2 in an increasingly complex military environment. As a multidisciplinary science associated with managing behavior of dynamic systems, control theory was one plausible means to improve the performance of the C^2 system.

Preliminary to management of dynamic systems, however, is the formulation of models that capture the system's essential behavioral characteristics. Looking back on the studies published as early as 1978, one sees a point-of-view that focuses primarily on the C^2 organization and process as the object of modeling and research, including models of the decision maker, the decision-making process, and even models of career progression along with models of the C^2 structure. Within the past few years, however, the perspective of the control science community has broadened to include C^2 in the context of battle management. To the eye of a control engineer, a commander is fundamentally trying to do control. That is, the military decision-maker is trying to achieve desirable behavior (from his/her perspective, not the adversary's) by judiciously committing resources and courses-of-action to the battlespace. Given this point of view, perhaps a commander's job could be made easier with the decision aid of formal control science and related methods.

Following this line of thought, the goals of the JFACC program were summarized by two key terms: *stability* and *agility*. The program hoped to show that control theoretic and related techniques have the potential to automate much of the C^2 decision process, thereby reducing the level of human tasking, and creating a system that rapidly responds to battlespace dynamics via incorporating feedback. Time-critical targeting is only one of many identified examples in which the *agility* of the C^2 process demonstrates military utility. So far as *stability* is concerned, one goal set for the researchers was to develop C^2 approaches in which plan-to-plan variability— that is, the amount of change needed to accommodate changing priorities, resources, or situation—was the least needed to maintain reasonable and acceptable behavior.

Approximately a year of intensive research had occurred along these lines when I took over from Col. McCorry as Program Manager. Quite frankly, I was initially skeptical about the extent to which meaningful contributions could be made to the military operational problem domain without significant maturation of the state of the art. Perhaps my main concern—and it remains a concern even now—is the low resolution of operational fidelity in the models and techniques the researchers were using. To apply the mathematical optimization algorithms that characterize modern control theory, many researchers were forced to simplify assumptions to achieve computational tractability. This simplification, in turn, raises concerns about the credibility of the results they produce, and hence of the results' ultimate military utility.

Having raised this large *caveat emptor*, let me point out some compensating factors that help to alleviate those concerns—factors that I believe create a case for seeing control science as a key enabling technology in future C^2 systems. First, a given technology must be embedded within a context, a concept of operations. As you will see repeatedly in the examples presented in this book, while too aggressive an application of these technologies is clearly ill advised, properly bounded use can provide significant advantage. Second, although some researchers have simplified assumptions for the sake of accelerating their investigations, they recognize that an operational system will require increased fidelity, and they have provided arguments regarding the scalability of their proposed approaches. Third, as the military moves toward increased automation and robotic/autonomous force structure, the modeling difficulties associated with the unpredictability of human action and intention will diminish, and the need for very rapid control decisions will increase. Despite the current modeling limitations, control science and other feedback-based technologies do appear to provide a path toward achieving those objectives.

The worked examples that constitute the bulk of this book grew out of an exercise within the JFACC program to assess the potential utility of these technologies by constructing well thought out and reasonably complete concepts of operation: an applicability study. The idea was to think through a potential application end to end, including such issues as human factors, infrastructure, and plausible and compelling scenarios. Although many of these examples have, for historical reasons, an air operations and often a specifically Air Force flavor, the ideas contained herein are service-neutral, and apply to a wide variety of military operations. It should be noted, however, that most of the technologies discussed in this book are as of yet exploratory efforts. The implementation of these technologies in realistic scenarios does not imply the existence of an off-the-shelf product—this text simply uses examples as a means to envision future operational utility.

A goal of this book is to present the potential military utility of control engineering in a way that the non-specialist can understand and evaluate. We chose to adopt a style that is neither a collection of technical papers nor an entirely non-technical discussion, resulting in a book that does not require a specialist's background and yet provides a significant measure of technical content. We avoid jargon and formalism, and focus instead on how the technology will work in practice—its limitations and dependencies, as well as its utility and value.

The task of leading the JFACC applicability study was assigned to the program system architect, headed by Dr. Alexander Kott. All members of the JFACC program, including the researchers, were actively involved both in the applicability study and in its distillation and reformulation into the publishable version that is before you. In this book, we begin with a historical overview of C^2, followed by a set of technology-focused chapters. Each technology chapter is organized approximately as follows: a vignette or an argument introduces the challenge addressed by the proposed technology, followed by a detailed review of the technology, next a scenario depicting the use of the technology, and finally a discussion of specific barriers that the technology would have to overcome—including human factors and system infrastructure issues. We then conclude with several chapters that discuss overarching issues common to any application of advanced technology to C^2, with remarks on future directions.

I want to acknowledge the hard work and insights of all who have contributed—researchers, architects, writers, reviewers, and editors. This book represents the efforts of many people with many points of view, and all have contributed to the quality of the final product.

June 1, 2001
Dr. Sharon A. Heise, Major USAF
Program Manager
DARPA/ITO

PART I
INTRODUCTION

CHAPTER 1

COMMAND IN WAR:
A HISTORICAL OVERVIEW

INTRODUCTION

In any attempt to understand the nature of command in war, which I use to include control as well, the first essential step is to separate it from leadership, with which it is often confused. Leadership consists of using various means—a mixture of example, persuasion, and coercion—to motivate and inspire people into doing what is necessary and what is expedient; to the contrary, command consists of using information to coordinate both people and resources in such a way as to carry out a mission. Whereas, in war, both leadership and command are absolutely indispensable. Of the two, the former is probably the more important and also the more difficult.

The need for command arises from, and varies with, the size, complexity, and differentiation of an armed force [1]. A one-man force does not require command, at least not in the sense that a 100-man force does. A force operating as a single, solid, homogeneous block (for example, a phalanx) would be, and is, comparatively simple to command—to the extent that it could be commanded at all. Once a force of any size is divided into several components, however, the problem of assigning a specific mission to each one, and of ensuring proper coordination among all, becomes much more complex. As the number of components, the power and range of their weapons, the speed at which they move, and the extent of the spaces over which they operate grow, then this growth will cause the degree of complexity to rise geometrically. Should the components in question become specialized—that is, acquire many different characteristics, each of which is tailored for a different kind of mission—then commanding and controlling them all will become a very complex business indeed. To turn the proposition around, the role of command grows with the size and sophistication of the armed force in question—which explains the amount of attention paid to it in recent years.

The responsibilities of command can be divided into two parts. First, it is necessary to order, arrange, and coordinate everything required for the day-to-day existence of a force, such as its manpower, logistics, medical services, and military justice apparatus. Second, it is necessary to enable the force to carry out its proper mission in war; to this part of command belongs intelligence-gathering, for example, as well as the planning, execution, and control of operations. Before 1945, the German and British command systems used to draw a very sharp distinction between the two parts. Since then, most forces have switched to the American system, which in turn is based on the French model, in which the two functions are regarded as co-equal and are integrated.

An equally useful way of looking at command is to ask not what it deals with but instead what it does. The answer then becomes clear that the exercise of command involves many things. There is, in the first place, the gathering of information on one's own forces, the enemy, and the environment. Once that information has been gathered, means must be found to communicate, store, retrieve, filter, classify, distribute, and display it. On the basis of information thus processed, an estimate of the situation must be formed. Detailed planning must be initiated. Orders must be drafted and transmitted, and their arrival and proper understanding by the recipients verified. Execution must be monitored with the aid of a feedback system, at which point the process repeats itself. Needless to say, in practice the various phases are likely to overlap.

Another interesting way of looking at command is to analyze the qualities of an imaginary ideal command system. An ideal command system, then, should be able to gather information accurately, continuously, comprehensively, and quickly. The system must develop reliable means to distinguish the true from the false, the relevant from the irrelevant, and the material from the immaterial. The system's information displays must be at once clear, detailed, and comprehensive. The mental matrix by which this information is turned into an estimate of the situation must correspond to the actual world, rather than to one that existed 25 years ago—or not at all. The objectives set by the ideal command system should be both desirable and achievable, two requirements that are often incompatible in practice. The alternative methods of action presented by the staff to their commander should be real, not just subterfuges presented as a matter of form. Once made, decisions must be adhered to in principle, but not under any circumstance and at any price; on the contrary, retaining flexibility and adapting to change *even as operations are going on* is a major requirement without which no command system can be successful. Orders should be clear and unambiguous, telling subordinates everything they need to know but nothing more. Once formulated, orders must be rapidly and reliably transmitted, and their arrival verified. Monitoring should be close enough to secure reliable execution, but not so close as to undermine authority and choke the initiative of subordinates at all levels. Finally, all this must be done securely; that is, without leaving the enemy any point of entry to learn what is going on, let alone presenting him with the kind of vulnerability that may enable him to influence the system or make it work in his own favor.

As even a cursory look at their nature will reveal, the functions of command are eternal. Provided he had a force of any size at his disposal, a stone-age chieftain would be confronted with every single one of them, just as is his present-day successor. In so far as the forces at the chieftain's command (and those of his enemy as well) were very much smaller, simpler, and slower, the functions of command were also simpler to carry out. However, our imaginary chieftain would hardly have writing materials, not to mention a pair of binoculars, to assist him in exercising those functions, which in turn would place very close limits on the size and capabilities of the forces he could command. Balancing the task facing him against the means available for carrying it out, there is no reason to believe that the exercise of command has become more difficult since Alexander showed how it should be done, for instance.

To command the forces at their disposal, present-day commanders have at their disposal something known as a command system. Definitions of systems vary, many of them being abstruse to the point of uselessness. In the present context, though, a command system may be said to consist of three distinct elements. These are organizations (such as staffs), procedures (such as the regular submission of evening reports, communication channels, distribution lists, and the like), and technical means (ranging from the javelin that Joshua's men waved to signal a move [2] all the way to modern computer networks). The combination of these three should make it possible to say a lot, though possibly not all, about the operation of any given command system belonging to any given force at any given time and place.

Just as it is possible to write a history of tactics, or a history of logistics, so command, too, has its own history, which is to some extent autonomous and independent of other elements of war. In particular, since command systems are made up of disparate elements, a development in any one of them will almost always entail changes in the rest as well. For example, once the procedure of submitting detailed daily strength reports became established early in the nineteenth century, specialized personnel for processing them also became necessary, giving birth to the first proper general staff. Once staffs existed, pen and paper, not to mention desks and filing cabinets (useful as status symbols as well as for practical purposes), became much more important than they had previously been. In other words, the various elements that comprise a command system interact with each other. By so doing, they push development along.

However, command systems do not exist in a vacuum. Their development is partly a response to changing requirements; obviously, a squadron of modern fighter aircraft, capable of flying half-way around the globe in less than a day, cannot be commanded by the same means as could a Roman legion, marching forward steadily at somewhat under 20 miles a day. To take the example of the Wehrmacht in World War II, a force numbering more than three million men, and attacking across a front 2,500 kilometers wide, cannot be commanded with the means used by Napoléon at the Battle of Waterloo in 1815. In short, command systems reflect the art of war as it exists at any given time and place. They are affected by, and in turn affect, that art.

For example, it was the introduction of the telescope that enabled a commander such as Frederick the Great to establish his headquarters at a fixed location overlooking the battlefield rather than rush all over it, as was the practice of Gustavus Adolphus a century earlier. Samuel Morse hardly had the U.S. Army in mind when he invented the telegraph; but once the telegraph existed, it did not take long for its military significance to be appreciated and, sometimes, exaggerated. The same was even more true of the telephone—originally invented as a by-product of research aimed at aiding the deaf—and the radio. The effect of all these inventions on organizations and procedures, indeed on the conduct of war itself, has been profound.

Following the above, this chapter is divided into five parts: (1) The Stone Age of Command, reaching from the earliest times to 1800; (2) The Revolution in Command, comprising the Napoléonic era; (3) The Age of Wire, from approximately 1850 to the end of World War I; (4) The Age of Wireless, from 1939 to 1991; and (5) Looking into the Future.

THE STONE AGE OF COMMAND

The Stone Age of Command lasted from the dawn of recorded

history to the end of the eighteenth century. Its dominant characteristic was the fact that, though changes in weaponry were numerous and sometimes frequent, the means used for the transmission of information changed hardly at all.

Under George Washington in 1780 AD, as under King David around 1200 BC, the most important means were acoustic and visual signals on the one hand, and messengers—either mounted or on foot—on the other. All of these suffered from various limitations. Musical instruments, such as drums and trumpets, and visual signals, such as flags and standards, could transmit information almost instantaneously; however, the amount of information they could convey was limited to a few pre-arranged signals, which might pass unnoticed in the din of battle. Their range, too, was strictly limited; while relays could be and were used, doing so was very expensive and compounded the likelihood of errors. Unlike acoustic and visual signals, messengers could carry information—oral, written, or both—over long distances. However, messengers tended to be slow and—especially if the enemy was close at hand—unreliable. Should a commander try to circumvent these limitations by making messengers work in relays or multiplying their numbers (duplication), then they, too, became very expensive.

What is more, neither reliable portable timekeepers nor large-scale two-dimensional maps—other than those based on guesswork—became available before the late eighteenth century. Thus, coordinating the movements of mobile forces over considerable distances was all but impossible. Except when opposition was expected to be negligible, as during the later stages of Alexander's campaigns in Bactria and Hyrcania, armies were compelled to march in single bodies until they met the enemy; who, on his part, was forced to proceed in a similar way. Marching in closely packed bodies—even in the eighteenth century fronts were seldom more than three to four miles long—armies were essentially limited to a few simple choices: advance or retreat, turn right or left, attack or defend. As Clausewitz notes, since armies were rarely able to force a battle, they issued challenges instead. In this sense, battle itself was a kind of ceremony or sporting encounter.

The inability of armies to coordinate the movements of separate bodies over large distances was reflected in the terminology. Any force operating more than a few miles away from headquarters could not be controlled and was, accordingly, known as a detachment. Strategy, here understood in the Clausewitizian sense of coordinating forces before and after the battle in order to accomplish the objective of the campaign, did not exist. First invented by Joly de Maizeroy, the term itself only entered the vocabulary during the years immediately before 1800.

While difficulties of C^2 prevented the rise of strategy as we understand it today, at the tactical level the available means permitted various solutions. The simplest one, much beloved by the Greeks and the Swiss after them, was to compress the troops into a single block—a phalanx—and launch them forward like some unguided missile. Thus, the coordination of all would be assured at the price of very strict limitations on the things they could do and the maneuvers they could carry out. Armies operating in this way did not really need a commander; with the result that he donned armor, took up a weapon, and fought along with everybody else.

Another method for coping with the limitations of C^2 was for the commander to take the lead over what he hoped would be the decisive wing while all but relinquishing control over the rest. For example, Hellenistic kings usually put themselves at the head of their armies' right wing, trying to force a decision while leaving their left and center to some trusted subordinate who was expected to operate more or less on his own. Gustavus Adolphus acted as a kind of mobile fire brigade to his own army, rushing to the aid of whatever formation seemed to require it and ending up by getting himself killed as he did so. As late as the beginning of the eighteenth century, the Duke of Marlborough could still be seen operating in this way. He moved from one wing of the army to the other; and, though he did no longer fight in person, came close enough to the front line for his life to be in very real danger.

Of all these methods, the most successful was the one adopted by Roman army. Like other armies, it had to reckon with the constraints imposed by the available technologies. Unlike other armies, it overcame those constraints by means of proper organization and proper procedures. Roman armies were made up of numbered, self-contained legions, which in turn were divided into cohorts, maniples, and centuries. Taking up the famous checkerboard formation, these units were spaced sufficiently far apart to avoid mutual interference but sufficiently close to permit mutual cooperation. Each one was commanded by an experienced officer. Each such officer had at his disposal a bugler and a standard-carrier to transmit his orders and to signal to neighboring units. The signals themselves were rehearsed almost daily and, during battle, carefully sequenced so as to make sure they did not escape the attention of the intended recipient [3].

In turn, this decentralized organization untied the hands of commanders. From the time of Scipio Africanus, the Romans would hold regular councils of war, in which the senior legionary commanders participated, and during which they would explain their intentions. On the day of battle, the senior commanders would take up a position on a hill in the rear. From there, conspicuously dressed in red togas, they would oversee the battle, only exercising loose control and only interfering when necessary. All this explains why, at such battles as Zama and Thermopylae, Magnesia and Cynocscaphlea, as well as the one waged by Caesar against the Belgians, we hear of tribunes and centurions who knew what to do or acted on the spur of the moment and won the day. It also explains how, even though Roman hardware was no better than that of its enemies, Roman software—in the form of organization and procedures—enabled the Romans to overcome the limits that their hardware imposed. It was to the point, indeed, that the Roman army became perhaps the greatest military force of all time.

THE REVOLUTION IN COMMAND

As so often happens, the changes that revolutionized C^2 during the early years of the nineteenth century did not come entirely out of the blue but had been foreshadowed by the operations of the French Army during the Seven Years War. As so often happens, too, in some ways the changes were the product of accident rather than design. The decisive factor was the proclamation of the *levee en masse,* which caused a mighty leap in the size of armies. Previously, few field armies had numbered more than 70,000 men, and the Marshal de Saxe at one point had written that no commander had it within his powers to command more than 50,000. Now, however, armies numbering 120,000 men and more became commonplace, both with the French and their enemies.

Such enormous armies could no longer march, let alone feed (supplies were obtained by requisitioning) along a single road. They had to be spread out over several, which meant that fronts grew from three-four miles under Frederick the Great to as much as 25-50 under Napoléon. Normally, the *Grande Armée* was divided into eight corps. Each numbered 20-30,000 troops, and each had its own permanent identity, commander, staff, combination of the three arms, and means required for communicating both with General Headquarters and the subordinate divisions. The corps' size and composition enabled each one to hold off even a superior enemy for a day or two, which in turn permitted General Headquarters to send the corps much farther away without having to fear for their fate. Contrary to most of the literature which presents Napoléon as a centralizer, what actually took place was an unprecedented *decentralization* of command from General Headquarters to the corps; at the same time, improved maps and timekeepers made it possible to coordinate the latter's moves.

Instead of simply marching towards each other, armies were now able to use one corps for fixing the enemy by demonstrating at his front. The second corps could outflank him on the right or left. The third cut off his lines and communications, the fourth guarded against enemy reinforcements, and the fifth acted as a general reserve. Moreover, these roles could be switched as fast as the necessary orders could be issued and the necessary messengers sent on their way. In this way, strategy, which for millennia had been tied down by the very limited capabilities of C^2, was liberated like Minerva springing from Jupiter's brow. And strategy's inventor, Napoléon Bonaparte, was put in a position where he could overrun a continent within a few brief years.

The factor that held together the marching, maneuvering corps— they have been compared to the waving tentacles of an octopus— was, of course, information. A vast stream of information, which had to be written down on special forms designed by Berthier, Napoléon's chief of staff, reached General Headquarters every day and was supplemented by even more detailed reports that had to be submitted every five and 15 days. Having arrived, the information was properly sorted, indexed, and catalogued before coming to the Emperor's attention. He thus possessed a constantly updated picture of what his forces were doing and what their situation was—an indispensable prerequisite for action that seems to have been much superior to anything available to his predecessors.

Napoléon, however, was well aware that information coming from the bottom tends to become less detailed and more stereotypical as it is passed from one echelon to the next. To get the information he needed at the time and place he needed it, and also in order to double check on what was coming from below and passed on by Berthier, he used a directed telescope consisting of two parts. At the highest level, he employed a number of brigadier generals sent out to reconnoiter entire theaters of war, monitor the movements of his marshals, and the like. At a lower level, so-called ordnance officers were sent to check on smaller units as well as observe (and sketch) fortresses, river-crossing points, and so on. This method enabled the Emperor to match carefully the information-seeking means at his disposal with the objectives in which he was interested; as he knew, there is no point in having a major check on a marshal, or a general on a captain.

Napoléon's system, and the results that it helped achieve, were remarkable in two ways. First, it was confined almost entirely to the strategic level; at the tactical level, Napoléon's contribution to the evolution of command was much more limited. Unlike Gustavus and Marlborough, Napoléon usually refrained from coming so close to the front as to put his life in danger, let alone take up a weapon and fight in person. Like Frederick the Great, he and his suite would take up a position on some hillside overlooking the field, changing position perhaps once or twice during the battle. Command was exercised by means of the telescope—the colorful uniforms of the day made it easy to identify individual units—and mounted messengers. Usually the latter did not proceed further down than corps headquarters, but on occasion they might be used to reposition and direct individual regiments. Though Napoléon was perhaps the greatest military genius who ever lived, even for him to command eight corps amidst the stress of battle could prove too much. At Jena in 1806, for example, he only oversaw the operations of four, forgetting about the rest, and operating in complete ignorance of the fact that one of them (Davout's) was fighting and winning a major battle against the Prussians. At Leipzig in 1813, where he had no fewer than 180,000 men, he only commanded one of the three battles that were going on simultaneously, while letting the other two take care of themselves.

The second, and even more remarkable, aspect of the Napoléonic revolution in command was the fact that it was achieved without any real technological innovation, let alone technological superiority over his enemies. With the exception of the optical telegraph—a relay system capable of transmitting information 200-250 miles a day—all the means Napoléon used had been around for centuries, if not millennia. In any case, since it was fixed in geographical space, the optical telegraph itself could not play a role in operational command. Indeed, Napoléon's revolution was achieved in spite of the fact that drums and trumpets were still nothing but drums and trumpets, flags and standards no better than before, and messengers and horses no faster or more reliable than they had ever been. In this way, Napoléon's achievement presents an object lesson in the fact that proper organization and proper procedures can revolutionize warfare, and can lead to very great victories even in the absence of technological change and/or superiority. So it has always been, and so it will always be.

THE AGE OF WIRE

If it is true that war is the father of invention, then surely communications technology has long been its unloved stepchild. As just noted, for millennia the art of war—organization, weapons, and tactics—was able to evolve without any corresponding advance in information technology. As a result, at the end of the eighteenth century armies were scarcely larger, the fronts over which they operated scarcely longer, and the repertoire of the strategic maneuvers they could carry out no more extensive, than had been the case 2,000 years previously. Napoléon, driven by the vastly increased numbers of troops at his disposal, instituted a new organization and new procedures, and was able to take existing technology to its limits. Within a few years, he revolutionized command and invented strategy; still, even with the best procedures and the best organization, there are limits to what any given technology can do.

This is hardly the place to elaborate on the origins of the telegraph, and its place as the most important innovation in communications technology since writing. Suffice it to say that experiments with the use of electricity for the transmission of messages had been going on since the 1760s, and that they reached fruition in the 1840s. In the 1850s, most European armies had started experimenting with the new instrument, assessing its advantages and weighing its limitations. The greatest advantage of the telegraph was the fact that it could deliver information incomparably faster than any other technical means previously designed, thus connecting forces operating at a very great distance with headquarters as well as with each other. The telegraph's greatest limitation consisted of a slow rate of transmission—although that was later to improve—and its dependence on wire. A wire-dependent system was slow and hard to establish, but easy to interrupt. It could therefore be expected to be more useful with large units rather than with smaller ones, on the defense rather than on the offense, in a stationary campaign rather than in a mobile one; and in one's own territory rather than the enemy's.

By way of illustration, the new instrument greatly influenced the 1866 Austrian-Prussian War. At the outset, the telegraph was heavily used for mobilization and deployment, given that it now connected most main cities and also ran along the railway lines. The telegraph permitted the Prussian forces to be grouped into three—later, two—large, self-contained armies. The armies themselves were strung out along a front over 200 miles long, causing Marx's friend, Friedrich Engels, who was a specialist on military affairs, to write that the King of Prussia had gone mad. Engels was proved wrong. Thanks to the wires that the armies laid out behind them, doctrine put the responsibility for doing so on them, instead of GHQ, it had become possible to carry out a maneuver on external lines far beyond anything Napoléon had ever attempted. All the while, the armies' movements were followed and monitored from Berlin.

While technical progress in the form of the telegraph constituted one leg of the Prussian advantage, the other, in the form of the General Staff, was no less important. Owing to Germany's controversial role in both World Wars, so much praise and opprobrium have been heaped on this institution that it has become almost impossible to discern its real contribution; in my view, it was as follows. First, the members of the staff, having gone through a common training at the *Kriesgakademie,* and knowing each other well (even in 1870-71 the entire staff, responsible for looking after a force over 300,000 strong, only counted some 70 officers), acted as a directed telescope. The staff did so, either while working for General Moltke in Berlin, with whom each army chief of staff maintained a channel of communication independent of the army commanders, or for the commander of each army separately. At both levels, the directed telescope cut through, and complemented, the standard reports coming in from below. Such communications thus provided senior Prussian commanders with an unrivaled capability to keep themselves informed on friendly forces, enemy ones, and the environment.

Second, and perhaps even more important, the staff acted as a data processing machine. The machine was made up of human parts; well organized in advance rather than the result of some hasty improvisation, the machine relied on a carefully worked out division of labor as well as fixed, well rehearsed procedures. These qualities enabled it, quietly and efficiently, to classify, evaluate, and organize the information that reached it from below, proceeding with a regularity and comprehensiveness far superior to anything in previous history. To put it in another way, the staff took over and systematized functions which, from the time of Alexander the Great until that of Napoléon and his opponents, had been carried out partly in the commander's head; partly with the aid of councils of war, consisting of senior subordinates who were assembled especially for the purpose; partly by the secretaries, who surrounded the commander and who might or might not have received special training for the task; and partly by various hangers-on, who, thanks to the emergence of the staff, could finally be dispensed with.

Between them, the telegraph and the General Staff enabled Moltke to control his forces from Berlin. Unlike previous commanders, he took the field only two weeks after his forces had first crossed the Austrian border, so as to be present at the Battle of Koenniggraetz. Unlike previous commanders, he commanded from behind his desk rather than from behind his horse's ears; not by accident did the expression estimate of the situation (implying map study) take the place of *coup d'oeil* (the commander's ability to take in terrain at a glance) during those years. Partly because the telegraph was still a slow and clumsy instrument, partly as a matter of deliberate and well considered policy, Moltke contented himself with brief directives and gave subordinates a degree of leeway rarely equaled in history before or since. Yet another reason for this leeway was that the telegraph, owing to its dependence on wire and the cumbersome nature of the apparatus it demanded, could do little to help in the exercise of tactical command. Communication between the armies, as well as with their subordinate corps and divisions, remained almost entirely dependent on messengers. Further down, too, commanders continued to use voice, drums, bugles, flags, runners, homing pigeons, and the occasional pre-arranged cannon shot serving as signals.

If anything, the problems of tactical command were growing worse. Not only were armies larger than ever—each of the three commanded by Moltke was the equivalent of several Napoléonic corps—but they operated in a much more dispersed manner. Dispersion, in turn, was dictated by the growing firepower of breach-loading, rifled small arms and artillery pieces. Regardless of what officers might do or say, the more intensive the fire, the more pronounced the tendency of the troops to break formation and take cover; the result was confusion and the kind of raggedness that characterized both Koenniggraetz and many a Civil War battle. Grasping the problem, Moltke was powerless to solve it. In a sense, what he did was to substitute strategic C^2—which, thanks to the telegraph and the general staff, were becoming more feasible—for tactical command, which, for the reasons just explained, was becoming more difficult. It worked: Koeniggraetz was decided by the movements orchestrated by the Prussian general staff, rather than by the fire of the troops on the ground.

By World War I, these problems still had not been solved. It is true that radio was available, but its usefulness in trench warfare was very limited, owing to mutual interference. It is also true that the telegraph had improved, permitting several messages to be sent over a single wire (thus increasing capacity) and reaching as far down as the regiment. However, telegraph's continued dependence on wire, and the difficulty of rapidly constructing lines in the field, still made it much more suitable for defense than for offense—thus helping explain the rise of trench warfare. At the tactical level, rapid-firing, recoil-less artillery, as well as smokeless powder, magazine-fed rifles, and machine guns, forced troops to take cover. At any level above the company, they were no longer visible to their commander's eye. On the attack, the only way to communicate with them—if they could be communicated with at all—consisted of pre-arranged signals, such as rockets, as well as runners braving the storm of steel.

In principle, there were two ways to solve the impasse. Both were tried at different times, by different belligerents, and with very different results. One was to impose order by force and attack according to a pre-conceived plan that had to be strictly adhered to. This was the solution adopted by the British at the Somme in 1916. Each infantry or artillery unit was assigned a lane perpendicular to the front. Each unit was told to march or fire across the lane, looking neither right or left, and proceeding at a pre-ordained pace. Advancing shoulder-to-shoulder, and forbidden to break into a run or take cover—doing so meant they could no longer be commanded—the infantrymen were to halt after reaching their objectives; for fear of coming under their own artillery fire, they were not to exploit success even if it were attained. Switching from target to target, as laid down in the pre-conceived timetables, the artillery could only change its plans by the authority of corps headquarters, normally located about seven to 10 miles behind the front. Perhaps worst of all, senior commanders, trusting in the wires, and afraid lest they would no longer be able to talk to their subordinates once the attack had started, actually prohibited officers from battalion commander up from joining their troops—a prohibition which, to their credit, some officers chose to ignore.

The upshot of this attempt to overcome the limitations of the available technology—by prescribing slavish adherence to a preconceived plan—is almost too well known to be repeated here. On the first day alone, the British Army suffered 60,000 casualties, including 20,000 dead, for no visible gain. General Douglas Haig, who had hoped to control the battle by the above means, ended up being the worst informed man in the entire Army. Came evening of July 1st, he did not even know that one of his corps had never left its trenches, let alone exercise control over it.

The second way to overcome the limitations of technology—this time, wire-bound technology—was once again found in the field of organization and procedure. The Prussian-German army has often been blamed for the corpse-like discipline (*Kadavergehorsam*) that allegedly prevailed in its ranks. In reality, and starting with the reforming generals Scharnhorst and Gneisenau in 1807-15, the army had developed a remarkable command system that put great emphasis on decentralization, individual initiative, and free cooperation between its members, especially the commander and his chief of staff. As originally conceived, the system was intended for the highest headquarters only. Later, though, dispersion and the growing emptiness of the battlefield caused *Auftragstaktik* (known in English as mission-type orders) to percolate down through the ranks. After 1871, it was company commanders, the highest grade, who could still expect to observe their subordinates and speak to them, who became the critical link in the chain of command. By 1908, the official regulations read: "war demands from every soldier, from the youngest recruit upwards, the total *independent* commitment of all physical and mental forces." Accordingly, as he prepared the spring 1918 offensive, General

Ludendorff did *not* follow Haig's example of trying to solve the difficulties of C^2 by prescribing a rigid plan.

Throughout the winter, subordinate units up to regimental level were systematically taken out of the line and retrained, so as to be able to operate *without* a plan. Each storm party consisted of a mixture of all arms—rifles, machine guns, flame-throwers, and light artillery (German artillery in 1918 was controlled by division rather than by corps; some of it was even put at the disposal of regimental commanders). Each storm party was assigned an objective and told to push forward as fast as possible, bypassing opposition, relying on its own judgment as to the best route to take, exploiting opportunities as they occurred, and cooperating with neighboring units as long as doing so did not endanger the pace of the advance. The artillery, instead of operating by predetermined timetable, was to shift its fire by observing the colored rockets launched by the infantry as it reached each successive line. The whole effort was to be coordinated by staff officers. Instead of being prohibited from leaving their telephones, the officers were sent far forward, roaming over the field, acting as the commander-in-chief's eyes, and using runners or pigeons to carry back the latest news.

The results of this attempt, too, are well known. Though the inability of supplies to follow up over the churned-up battlefield caused the Germans, too, to run out of momentum, it took them only four days and relatively few casualties to recapture ground over which the British in 1916 had struggled for four months. A unit-by-unit analysis of the battle shows that the success of each unit was directly related to the extent to which it applied, or did not apply, Ludendorff's principles. What is more, as the offensives of May and July 1918 indicated, it was a triumph capable of being repeated.

The lesson of these events for the art of command in war is obvious. Given any level of technology—and the technology used by both sides in 1914-1918 was remarkably similar—there are two ways to use it. One is to limit oneself to whatever the technology can do; the other is to gain a thorough understanding of the things that technology can *not* do and then rely on organization and procedure to do them nevertheless. Otherwise put, and regardless of the technology in use, the Roman legions, Napoléon's *Grande Armée*, Moltke's armies in 1866 (and 1870-71), and Ludendorff's storm troops all had this in common: that they did not wait for detailed C² from above. To the contrary, in all four cases the whole idea was to organize and train the armies in such a way that they should be capable of operating even when that command and that control broke down. As Clausewitz wrote, war is the most confusing and confused human activity of all. And, in war, he who can thrive on chaos wins.

THE AGE OF WIRELESS

As this chapter has shown, the history of command can be understood in terms of the distance separating commanders-in-chief from the forces under their command. Originally, the small size of armies probably dictated that commanders should fight in the front ranks. With the exception of the Roman legions between about 200 BC and 250 AD—the period when infantry began to disappear and was replaced by cavalry—this was also true of the ancient period. Medieval and even early modern commanders fought as a matter of course, and were often killed, injured, or captured as were Richard Lionheart and Francis I of France. In this connection, the fate of Gustavus has already been mentioned. As late as 1675, the French commander Turenne was killed when helping some of his men re-site a battery during the battle of Breisach.

Though there were exceptions—at Leipzig in 1813, General Duroc was killed by an artillery round while standing on a hill next to his Emperor—eighteenth-century commanders, thanks to the telescope, were usually able to command from a safe distance.

During the nineteenth century that distance continued to grow, until, in 1914, Moltke the Younger tried to command the invasion of France from Luxembourg, dozens of miles behind the front and as much as 150 miles away from his own right and left wings. With the telegraph lines unable to follow the advance, and German radio communications being interfered with by a powerful French transmitter (put on the top of the Eiffel Tower for the purpose), it is no wonder that Moltke lost control. In the end, he was forced to send a lieutenant-colonel, Hentsch, to motor right across the front so as to find out what was happening. By the time he did find out, the First Army had changed direction, the Second Army had begun to retreat, and the Battle of the Marne had essentially been lost.

The Age of Wireless reversed this process. Moving at the speed of light, radio waves easily overcame the handicaps of distance, terrain, movement, and weather that had often bedeviled communications in the past; in theory, at any rate, radio enabled everybody to communicate with everybody else all of the time. This was true even though operations were now conducted with the aid of armored and motorized divisions capable of advancing at perhaps 150 miles per day, even though they involved squadrons of aircraft flying at several hundred miles an hour, and even though battles were now so large that they could cover entire countries. Napoléon could spend an entire day standing on a hilltop near Jena while blissfully unaware that, 10 miles away, a third of his army was fighting the Prussian main force. One hundred-thirty-eight years later, General Marshal, sitting at his office in Washington, D.C., could have followed the Battle of Normandy by the hour, had he wanted to.

Such enhanced power—including the power to meddle and to interfere—was not without its dangers. Based as it is on fast-moving, far-ranging operations by armored divisions and aircraft, modern warfare is extraordinarily fluid. Opportunities are lost almost as fast as they are created; very often, waiting for the situation to be explained to higher commanders, so that they can make a decision, is to risk a setback or worse. Put into the hands of cautious commanders, such as Montgomery and Eisenhower, radio could present a positive danger and lead to a slow, deliberate style of warfare being imposed and controlled from above. Worse still, radio permitted high-ranking commanders to bypass or neutralize their own immediate subordinates and personally assume control over the operations of very small units hundreds of miles away.

An object lesson—as to what could happen when radio is misused—was given by the Israelis on October 8, 1973, at the height of the so-called Yom Kippur War. Ordinarily, the Israelis were very good at armored warfare, carefully balancing control with independence, and priding themselves on their junior commanders' initiative and willingness to act on their own. On that day, however, various considerations induced the chief of staff, General Elazar, to insist that only such moves be carried out as had received his prior approval. As it happened, communications on that day were bad, owing to Egyptian interference. To comply with Elazar's orders, the front commander had to position himself in a place where he could communicate with General Headquarters in Tel Aviv. Under him, division, brigade, and even battalion commanders were forced to do the same. As had happened at the Somme in 1916, each one did what he could to maintain contact with his superior, even if it meant he could no longer communicate with his subordinates. The result was disastrous, as two armored battalions were practically destroyed, and the commander of one was taken prisoner [4].

In the hands of a Guderian, a Rommel, or a Patton, however, radio served as an extremely powerful tool of command. By allowing communications with rear headquarters to be maintained at all times, radio enabled these commanders to separate themselves from their staffs, go far forward, and see for themselves. By eliminating the problems associated with distance, terrain, and the relative movements of units, radio made possible a combination of flexibility *and* control, which in turn was behind some of the greatest triumphs in the whole of modern warfare. To focus on the example of Patton in 1944-45, a detailed investigation has shown that his Third Army made less frequent use of top priority communications than did any other headquarters; to put it crudely, its commander refused to sweat. In turn, deliberately abstaining from frequent use of top-priority channels left them free for when they were really needed and gave everybody a feeling of confidence. Patton himself started the day by carefully briefing his staff as to what he expected to happen and what he wanted them to do. Next, he would mount his jeep—sometimes, a light aircraft—and spend his time touring the front. He visited units down to regimental level, intervened on the spot where necessary, and raised morale by means that were often as unorthodox as they were effective.

Far from conforming to his image as a gung-ho commander, Patton's method of C^2 was based on careful organization. He did, of course, rely on the regular situation reports passed on by subordinate units four times a day. In addition, though, he also had the so-called Household Cavalry. This was a unit whose sole function was to serve as a directed telescope by listening in on the networks operated by the units making up Third Army; significantly, though, its members were forbidden either to interfere in what was going on or to criticize it. Thus, Patton had at his disposal two sources of information. Each operated in a different way, and each could be used to check on the other. All available information was fed to rear headquarters, where most of it was taken care of without any intervention on Patton's part. Meanwhile, he himself would go to wherever the center of gravity was, speak to commanders, study the situation on the ground, and, on occasion, do what had to be done.

In 1991 at the Gulf, wireless still remained the most important single means by which operations were commanded. To be sure, a situation had long been created where no combat vehicle was without at least one radio apparatus, and where many carried several different ones. On top of this situation, technical means available to higher headquarters were often capable of transmitting not just voice but data and TV pictures. For reconnaissance and surveillance purposes, both of the enemy and of one's own forces, light aircraft and helicopters had been joined by satellites (which were very important) and RPVs (which were less so). Soon every soldier would have his hand-held, satellite-linked apparatus, giving him not just his exact coordinates but also enabling him to communicate with all of his fellow soldiers by voice, e-mail, or both. As had already begun to happen in Vietnam, in many cases the lack of information that had so often obstructed past commanders was turned into its opposite, a tremendous and almost unmanageable glut. For example, merely to catalog the millions of satellite pictures of the Kuwaiti theater of war, and to decide which of them to send to Marine Corps Headquarters in the Gulf, proved all but impossible. As a result, General Boomer's divisions ended up by attacking with hardly any information at all.

Both on the ground and in the air, a modern war of rapid movement consists of nothing so much as fleeting opportunities that must be seized and exploited. Given the available means of C^2, in this particular war some opportunities were taken, others lost. While there were no outstanding failures—except, perhaps, failure to realize, on time, that the Iraqi National Guard was getting away— neither can one say that there were outstanding successes. Perhaps our conclusion should be that, given the opponent's unexpected weakness, the system of C^2 did what it was supposed to do without, however, really being put to the test. This statement recognizes the undeniable accomplishments of the past. At the same time it serves, as it is intended to, as a warning concerning the future.

LOOKING INTO THE FUTURE

From Plato to NATO, the history of command consists very largely of an attempt to devise better communications-technology and use it in order to obtain certainty. Certainty concerning the enemy's capabilities and intentions; certainty concerning the many factors that together constitute the environment, from the weather and the terrain to the presence in the atmosphere of chemical warfare agents; and, last but not least, certainty concerning the capabilities and intentions of one's own forces, regardless of where they may be or the activities in which they may be engaged.

Also, on the whole, the attempt to use communications-technology in order to achieve certainty has only been a qualified success. Though vast advances have been made in the speed of communications, their range, their flexibility, and their capacity, the fog of war has probably not diminished at all. Consequently, at each historical stage, the best armed forces were those which understood how to use organization and procedure to overcome the limits of technology; in other words, operate in such a way as to take uncertainty in stride, even revel in it.

As to the future, two views exist. One, whose most important proponent is Admiral Bill Owen (Vice Chairman of the Joint Chiefs of Staff, 1994-95), sees the future in terms of using increasingly powerful computers to link together increasingly numerous, and increasingly sophisticated, sensors and information-gathering systems [5]. Many of the systems in question already exist; others remain to be developed. If enough money is spent, and everything works as it should, then in theory the result should be the so-called system of systems. Such a system should be capable of seeing and registering everything inside a theater of operations measuring 200 by 200 miles. After thousands of years during which commanders were often like blind men groping in the dark, and not seldom owing what victories they won to accident rather than design, the fog of war will finally be lifted. Command itself will be reduced to a matter of cool calculation as to what weapon systems should be assigned to service what targets, in what numbers, and in what order, with the aim of obtaining the best possible results with the least possible expenditure.

The other, and perhaps more realistic, vision expects uncertainty to increase rather than decrease [6]. In part, this vision is because nobody knows which technologies will be available at what time and whether they will work; in part, because of the growing complexity of war itself. Assuming this will be the case, and also to save casualties, it is intended to make a revolutionary departure. Hitherto all attempts to improve command have centered on providing better communications, better organization, and better procedures; now, the point has come where some military operations may perhaps be entrusted to robots with their own built-in decision-making mechanisms. No more than Roman legionaries needed the commander-in-chief's orders to know what has to be done in the midst of battle, will such robots be commanded from the center. On the contrary, the objective of the exercise should be to make them capable of operating on their own *without* having to be controlled in detail by that center [7].

Though some robots capable of making their way autonomously on the ground are currently being experimented with, the first operational ones are expected to be airborne. Devising a command system for uninhabited vehicles (UVs) represents a daunting challenge, one to which historical command systems, which have always consisted essentially of people and the procedures that they employ, can only provide very limited guidance. The first step will presumably consist of developing a reliable model to distinguish that which must still be done by humans from that which can be represented in the form of algorithms and left to the machines themselves. Next, the things that must be controlled by the center—for example, setting the mission—must be separated from the operational details which on-board computers can understand and act upon. The latter might perhaps include such things as deciding which target to attack first, selecting the most promising route to it, and activating countermeasures. Since on-board computing capacity may well be limited, one can envisage a middle tier of computers coming between the robots and headquarters.

To link the various elements, a system of reliable communications must be developed. This is no mean task, given the tendency of complex environments to interfere with wireless communications, and the near certainty that attempts will be made to jam or spoof them. Even as the mission is being carried out, the capability to change or abort it must remain; whether this capability is reserved for the center, or exercised by the machines themselves, it assumes the availability of additional information and, very likely, some kind of directed telescope. Assuming that more than one robot is used simultaneously (current experiments aim to show that one battle manager can control four UVs), even greater is the difficulty of making them communicate with each other. Let alone building them in such a way that they can share the information that they themselves gather and/or that is transmitted to them from the center; it is for this purpose, indeed, that an intermediary level may be most useful.

At present, whether a command system that can meet these and other requirements can be built is anybody's guess. Doing so represents a formidable challenge, but one which, given the direction in which future warfare is likely to move (a combination of asymmetric warfare waged in very complex environments with a very great reluctance on the U.S. side to tolerate casualties), may be the only way to proceed. On the plus side, this is 2001. The U.S., as the sole remaining Superpower, is as secure against massive military attack as it has ever been in history. With both control theory and the computer technology to serve it developing at an unprecedented pace, now is the time to experiment. In the words of Francis Bacon, this is because knowledge can only be attained by experiments, and because knowledge itself is power; *scientia, ipsa vis est.*

SUMMARY

Developments in the three distinct elements of a command system—organizations, procedures, and technical means—define distinct ages in the history of C^2.

The Stone Age of Command, from earliest times to 1800, is characterized by limited communication technologies—acoustic and visual signals, and messengers. Armies marched in single bodies, and strategy hardly existed. At the tactical level, command solutions ranged from a phalanx to Roman decentralized command procedures.

The Napoléonic Revolution in Command was driven largely by an increase in army size. Innovations included an unprecedented decentralization of command, the birth of strategy, vast flows of information, and supporting organizations and procedures.

The Age of Wire, from roughly 1850 through the end of World War I, was engendered by the invention of the telegraph, which delivered information incomparably faster and farther, and by the creation of the General Staff, which acted as a data-processing center for increasing information flows. At the same time, reliance on wire made tactical C^2 more difficult, leading to trench warfare and an over-reliance on rigid, pre-conceived plans.

In the Age of Wireless, from 1939 to the present, radio waves overcame the handicaps of distance, terrain, and movement, making possible a combination of flexibility and control. Radio also brought the power to meddle, and a glut of information.

As to the future, one view envisions systems capable of seeing everything—and reducing command to a matter of careful calculations. Another view holds that, historically, in spite of advances in communications, the fog of war has not diminished. At each historical stage, the best armed forces operated in a decentralized manner, taking uncertainty in stride. In the future, military operations may perhaps be entrusted to autonomous robots with built-in decentralized decision-making mechanisms.

CHAPTER 2

CONTROL IN WAR:
A TUTORIAL INTRODUCTION
TO CONTROL THEORY

If the future of C^2 indeed involves introducing automation into the process of command, as well as extensively using autonomic forces, one is justified to explore those scientific foundations that might make such advances possible. This book concerns itself with concepts formulated around applications of control-theoretic and related approaches to actions in a military battlespace. Thus, we felt that a brief tutorial on control theory, with a particular emphasis on how it might apply to C^2, would be appropriate at this point.

Control theory is notorious for its high content of abstract mathematics. Despite an intimidating reputation, however, many of the key ideas underlying the military applications of modern control theory have a strong intuitive conceptual basis and can be expressed without use of formal notation. Indeed, a few powerful analogies can suffice to pose a problem and its most important characteristics in a way that non-specialists can grasp. Although a *complete* solution will eventually carry a researcher into deep mathematical thickets, a broad understanding of any important issues and techniques lies conveniently near the surface.

Additionally, while control theory is a very large and active area of pure and applied research, a complete survey of control theory, even at a very cursory level, is far beyond this book's scope. Instead, this chapter is intended to prepare readers for material covered in subsequent chapters. As such, this chapter provides an introduction to, and an overview of, the application of control theory to military operations.

There are four major areas to discuss:

1. Control theory in a military context
2. Model Predictive Control (MPC)
3. *Game Theory* dealing with adversarial intent
4. Ancillary approaches

How should we model military operations so that control theory can be applied? There is no doubt that this is *the* over-arching question faced by researchers in this field. As will be shown in subsequent chapters, a large range of answers and approaches are possible. We begin with concepts common to many research approaches, therefore enabling us to pose controls problem for military operations suitably.

CONTROLLERS AND PLANTS DEFINED

Figure 2-1 shows a classic plant/controller block diagram, with the term *plant* borrowed from early control theory, when automatic monitoring and process management in a chemical plant was a major application. Here the *plant* signifies any ongoing process to which control theory can be applied to adjust or direct its behavior. As such, the *controller* is the algorithm which accepts observations about the state of the plant and makes decisions about what control signals to apply. As the figure shows, two types of data move between the plant and the controller: observations about the current state of the plant; and control signals issued by the controller to entities in the plant.

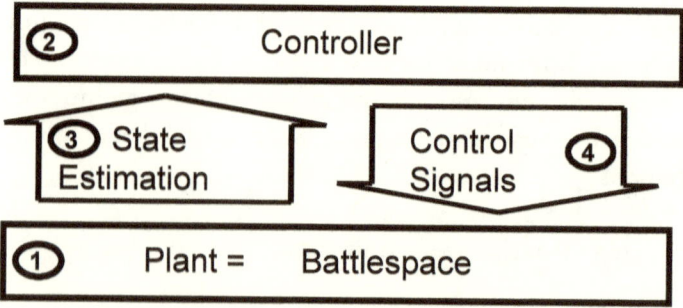

Figure 2-1. The controller/plant diagram represents an ongoing, continuous process of moving plant state estimation data into the controller, in order for the controller to issue or update control signals to entities in the plant.

As described in Figure 2-1, the information on the state of the plant flows more or less continuously and asynchronously into the controller, and the controller continuously (or as-needed) issues or updates control signals to entities in the plant. It is interesting to note, however, that such a fluid process contrasts with the rigidly sequential and synchronous OODA (observe-orient-decide-act) loop so well known in military doctrine.

Looking at the four blocks in turn, the plant in block 1 represents the portion of the battle to be controlled. As a single military illustration, consider the air operations of an overall campaign. On a purely *tactical* level, the plant might encompass the flight path and actions of a fighter, or a flight of aircraft, during a single attack sortie. The entities that would therefore make up the plant are the aircraft (friendly and hostile), surface-to-air missile (SAM) units, supporting resources (bases, refuelers, jammers, high-speed anti-radiation missile (HARM) shooters), targets to attack, and so on. The time period is a few tens of minutes. And the pace of activity (that is, the rate at which significant changes occur, requiring updates to and by the controller) is several times per minute.

If the problem is at the *operational* level, the entities are very different. Entities might consist of entire squadrons or air bases as well as models for the hostile array of battle; control signals would allocate resources, such as air tasking orders; and the time frame and pace could be up to several hours. Finally, at the *strategic* level, the abstraction is greater still, with a time horizon extending up to several days, and with a correspondingly aggregated (or summarized) set of reports and commands. All three of these levels— tactical, operational, and strategic—are included in the approaches that are discussed later in this book. Each, in turn, will provide heuristic cues for what entities are being modeled, and at what level of fidelity or abstraction.

In block 2, the controller will require a set of *models* of the plant; indeed, models are the *sine qua non* for a controller and are a major subject of study and research in their own right. They must be *formal models*; that is, they must contain enough rigorous structure to be amenable to mathematical analysis and optimization. Examples of formal models would include differential or difference equations (linear or nonlinear), Markov chains, probability models for noise or latency, models for target and asset value (perhaps time varying), probabilistic engagement and attrition models (for example, Lanchester equations), and environmental models (weather, terrain, infrastructure).

From the point of view of control theory, two key aspects of the model predominate. First, there is the *state* of the plant—that is, a formal representation of the important characteristics of the plant entities at any given point in time. Second, there are the *dynamics* of the plant—that is, the rules or laws under which the state will evolve over time. In addition to these models, the controller will contain an *optimization algorithm* to decide what control signals to issue to the plant. Typically, this operation involves *prediction* based on current estimated states and dynamics models. Basically, the controller will estimate how the plant can be expected to evolve when one or another of the available control signals is selected. The controller can then select the control that best achieves the commander's intent—which, in turn, is based on such factors as total (or expected) value of targets destroyed, value of assets lost, and probability and timeliness of achieving a desirable end state.

The task of optimization is complicated by many factors, including uncertainty about numerous entities in the battlespace (for example, lack of timely military intelligence reports as well as adversarial deception); uncontrollability (such as, enemy resources, weather); an active, intelligent adversary; and mismatches between behavior predicted by the model and the true behavior of the plant.

Per block 3, if the controller is to receive regular updates about the state of the plant, there must be sensing devices and a supporting infrastructure to collect, process, and distribute the information. In control theory, this flow of information from the plant to the controller is called *state estimation*, and it is a major field of study in its own right. However, the controller does not need all the information about the plant. In fact, only the information mathematically required by the controller's internal models and algorithms need pass across this interface.

While such limited information is a far cry from the notion of broad situation awareness commonly preferred by military C^2, nevertheless an advantage of such a formal, control theoretic approach is that the type, frequency, and accuracy of the required observations can be engineered from the controller's models and algorithms. Therefore, under such conditions levying defensible military intelligence requirements—where defensible means based on quantitative controller performance tradeoffs—is a realistic goal. In particular, control theory is often able to bound the performance of an algorithm as a function of the timeliness and accuracy of its data input. Of course, degrees of aggregation or summarization, and frequencies of input, will vary according to the type of plant—whether a tactical, operational, or strategic level is being controlled.

Finally, in block 4, the control signals issued to the plant will depend on the type of model (tactical, operational, strategic), as will their frequency and period of effectiveness. Typically, the farther up in the command hierarchy, the longer the duration or time span of the control signal. By contrast, at the tactical level, control signals may be issued and updated on a very frequent basis in response to the highly dynamic nature of the live battlespace.

A plan—a control signal that extends over a fairly long period of time—may be modified as new information becomes available, generating new or modified versions. Indeed, much of the research in this book has adopted this approach: issue a plan of action, then, in response to new information, modify it appropriately. Managing plan-to-plan variability, by measuring the amount of change needed to maintain satisfactory behavior, recapitulates the classic tension between agility (rapid response to new information or opportunities) and stability (firm commitment to previously issued command objectives). During the course of this book, the researchers continually wrestle with this tension.

INCLUDING THE ADVERSARY

One of the most challenging aspects of using control theoretic approaches in a military context is the presence of an active, intelligent adversary. While it is possible to model the actions of an adversary as a random disturbance (the same way, for example, weather is often modeled), such an approach may result in losing some of the most important characteristics of the system behavior, which in turn may lead to unacceptable results. For this reason, many researchers have chosen to extend the ideas shown in Figure 2-1 to include the adversary as an explicit component, as shown in Figure 2-2.

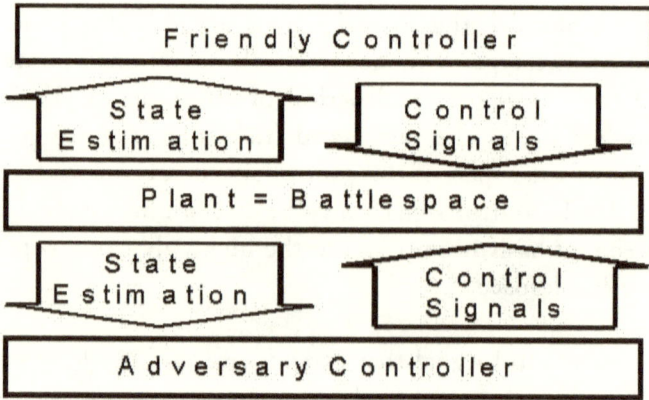

Figure 2-2. Modeling the adversary addresses the all-important issue of predictability—the element of surprise; doing what the enemy least expects—and certainly has great military utility.

Here, the adversary is treated fully symmetrically. Just as the model of the Blue force uses an optimizing controller, objectives, state estimation, resources, and controllable entities, so does the Red force model. Indeed, by extending the model to include the adversary, it becomes possible to estimate what the enemy's objectives might very well be. In fact, given all the data available, a good initial guess might be that his objectives are opposite one's own; if nothing else, that would certainly be a conservative, worst-case approach. In any event, using such an estimate one can at least calculate the control signals an adversary *might* issue (assuming, of course, that his decisions are rationally optimized). Then, using the models, one can test that hypothesis about the adversary and his intent against Red actions actually observed in the battlespace.

Using this model, researchers have to play both sides of the conflict in order to produce adequate and convincing results. Here, they cannot assume a passive or random adversary. Much to the contrary, to indicate the value of their approaches they have to assume a worthy opponent; that is, an opponent with optimizing control capabilities at least equal to their own. And finding control approaches that perform well in the face of an active, intelligent adversary is a hard task indeed!

With such an adversary there arises the all-important issue of predictability. The element of surprise—doing what the enemy least expects—certainly has great military utility. So by mechanizing or automating the planning process, do we not eliminate the element of surprise, hence making ourselves vulnerable to a more flexible and creative adversary? Just as we can try to estimate what an adversary would do, so an adversary can calculate what actions we might take, and pro-actively guard against or frustrate them. If such an argument were true, then using automated optimizing controllers could result in defeat, or reduced effectiveness, since we might become all-too-predictable to our adversaries.

While this objection has validity, there are two cogent reasons to employ modeling: rapidity of response and deterrence. Indeed, one vital reason to introduce control theory into military operations is to deal with time delays currently associated with solely human decision making. Just as the element of surprise has measurable, demonstrable military value, so has an ability to adjust to changing circumstances rapidly—and to exploit fleeting opportunities. After all, quick and predictable is presumably better than slow and unpredictable!

Further, by explicitly modeling the adversary's control capabilities—including the errors and latencies in his military intelligence capabilities and chain of command—we have the ability to gauge more accurately the various courses of action's probability of success. Thus, what we lose in predictability we more than recover in agility and predictive accuracy.

Then there is deterrence—creating in an adversary's mind the certainty, or probability, of defeat. If an adversary knows that we have deployed a C^2 infrastructure capable of *predictably* anticipating and rapidly responding to all reasonable courses of action he might take, this may be sufficient to dissuade him from undertaking the engagement in the first place. While such an argument may be less plausible on the strategic level, it is highly plausible at the tactical level, where a decision-maker is directly and personally affected by the consequences of his decision.

OBJECTIVE FUNCTIONS AND OPTIMAL CONTROL

For an algorithm to decide among various alternatives, it must have a mechanism for comparing them based on a calculation. In our situation, this calculation amounts to a prediction about how the plant, or battlespace, will evolve in response to one or another possible control strategies. In effect, the algorithm looks into its crystal ball to see what the future might look like for each of the options it is considering, and then selects the option that leads to the best, or most desirable, future state.

What is best? How is the algorithm to trade off, say, friendly versus enemy attrition? Rapidity of result versus friendly losses? Probability of success on front A versus success on front B? How is it to decide which, among available attack targets, should receive highest priority? And so on. Somehow, the algorithm must be able to balance these competing objectives computationally in a way that accurately reflects the human commander's intent. Ultimately, therefore, what makes an algorithm good is the extent to which it accurately implements the goals of the human being it is created to serve.

What this means is that there is a necessary translation process. On one side is a human being who speaks and reasons using natural language, and who, based on years of experience, has a heuristic understanding of the battlespace. On the other side is a computer, which operates in terms of mathematics and the chip-level logic of and-gates and transistor voltages.

Setting Weights, Making Priorities

From the computer's point of view, the easiest way to bridge this gap is through the use of *weights* or *priorities*. The algorithm designer constructs an objective function which, at its simplest, is a sum of the commander's key factors: the value of friendly assets and enemy targets (singly, or in combination, and indexed by time), the probability of achieving a well-defined objective, the expected length of time to achieve the objective, and so on. Attached to all these terms are weights (or priorities) intended to reflect each one's relative importance. By choosing a large relative weight for a particular factor, a commander indicates that he considers solutions favoring this factor better than solutions that don't. Conversely, by choosing a low relative weight a commander de-emphasizes the importance of a particular outcome.

With such objective functions specified, the algorithm then proceeds mechanically, producing a larger value for the objective function preferable to those that produce a lower value. In its work, the algorithm can exploit whatever mathematical structure is available—in the state and dynamics models, and the objective function—to produce optimized controls rapidly. While that, at least, is the conceptual ideal, reality is often a bit messier. What can go wrong?

First, it is a difficult matter to formulate objective functions that actually do what they are supposed to do. For example, it may not be clear which measurable (or estimable) properties should be chosen for inclusion. Often, surrogate metrics are employed when true conditions are either not measurable or not compatible with the mathematical modeling approach employed.

Even more troubling, however, is the burden placed on a decision-maker to assign values to each of the weighting factors. While an algorithm merely considers weights as input from the user, the user, or commander, must precisely express his intent when selecting the weights. If he wants to be more aggressive, he must express it through adjusting the appropriate terms' weights. If he wants to reduce friendly attrition, for example, even at the expense of lengthening the conflict, he must capture that by adjusting the appropriate weights. This, then, is the way the commander talks to the algorithm, by adjusting the weights entered into the objective function.

The operational significance of weighting cannot be overstated. How are the weights to be assigned? By whom? Can they be pre-assigned by preparing a number of advance templates to capture major known command stances or battle phases? Can they express complex, nonlinear dependencies and time sensitivities? Some researchers have suggested approaches for simplifying—or partially automating the weight assignment process. In any case, in examining the approaches to be presented, keep the objective function's existence in mind, and consider the operational impact of its initialization and maintenance.

Being Sensitive

Various calculations can help the weighting process, not the least of which is a sensitivity analysis. Here, taking into account the current state of the battlespace, the controller computes which available factors (and associated weights) will have the most impact on the overall objective function. This way, the algorithm can indicate to the commander which factors are currently most important. As a battle progresses, such factors will naturally change, since they depend not merely on the commander's intent but also on the battlespace and its entities.

Such an ability to perform sensitivity analyses gives rise to a new concept of operations in which the controller technology is employed not so much to issue commands, but rather to assess sensitivities. In effect, the controller automatically explores the high-dimensional space associated with the control of military operations, drawing the commander's attention to those factors that are currently most important. In this way, the controller acts as a kind of cognitive prosthesis, enabling the commander to act rapidly and intuitively in a regime whose complexity far exceeds unaided human capabilities.

Indeed, this type of cognitive computation acts as a value landscape, in which the objective function is imagined as sitting above the space that represents the plant. From a C^2 point of view, peaks, and areas with steep gradients, are where the proverbial action is. By contrast, flat areas represent factors with little or no likely impact on the course of the engagement—as in Figure 2-6.

FEASIBILITY: DEALING WITH CONSTRAINTS

Needless to say, solutions produced by the controller must be *feasible*—they must be capable of being carried out by entities in the battlespace. The limitations to feasibility, or constraints, can be captured in dynamic plant parameters, including maximum aircraft speed, probability of radar detection (as a function of distance and altitude), weapon lethality against various targets, and so on. Since they tend to be independent of any particular battle or engagement, such constraints can often be modeled and specified in advance. Other constraints, however, are highly situation-specific, and in fact can themselves vary greatly over the course of the battle—logistics constraints (availability of weapons and personnel), weather, military intelligence resources, terrain, transportation infrastructure, and so on.

Another constraint—more subtle, but in the end just as important—is the human acceptability of such advanced technologies. It is very likely that the optimal C^2 organizational structure—into which such visionary approaches would be employed—may differ, perhaps significantly, from current practice. Indeed, one can predict some resistance to such change, with the burden of proof on the innovators to show that the added value is sufficient to justify some painful organizational restructuring. New chains of command, and new staff positions with new job descriptions, will in all likelihood be required—while at the same time, some existing positions may be eliminated. Discussions regarding changes in concepts of operation will naturally arise during examinations of proposed new technologies. In such context rich settings, a technology's military value will be clearly apparent, as will the organizational infrastructure needed to support it. A military professional will be well positioned to assess the relative advantages and disadvantages of what is being proposed.

Another constraint with which we have not dealt in any detail is cost. As is the case in any field, such emerging technologies are in competition for scarce R&D dollars. Further, the technologies' implementation may require additional funding for military intelligence, communications, and computational infrastructure. Certainly, there might be associated savings, including reduced personnel and more efficient use of available resources. In any case, such possible quantitative cost-benefit tradeoffs will involve complex measurements and close observation of each particular usage. However, one of the advantages of the formal approaches espoused here is that they enable quantitative performance tradeoffs. For example, it is possible to show, and measure, the algorithm's performance degradation due to the latency of the data flows, the inaccuracies of the intelligence reports, or the overflow of the communications channels. This enables policy makers to quantitatively balance desired levels of performance against the increased resources needed to reach those performance goals.

We therefore focus exclusively on this purely formal approach—which leads to tangible results. For example, one can say precisely how an algorithm's performance will degrade according to reduced timeliness and lower military intelligence accuracy. Then such formulae will easily enable policy makers to create accurate performance levels and deduce necessary requirements to support such goals.

DUPLICATING THE REAL WORLD: TRADE-OFFS IN MODELING FIDELITY

The control-theoretic technologies described in this book rely crucially and inescapably on formal models—that is, models susceptible to mathematical analysis. A military problem is cast into a model susceptible to translation into mathematical language upon which, using the enormous power of today's computers, large and complex computations are performed. The computational result, we believe, has significant value.

Even so, given the extraordinary advances in computational power made possible by high-performance microprocessors and large-scale parallelism, not all models are equally amenable to mathematical analysis and numerical computing. Since we cannot hope to duplicate the real world exactly, the question of modeling fidelity inevitably arises. What has been abstracted away? What approximations have been included? What has been ignored entirely? These questions exist not only for the internal predictive models used by the controllers, but also for the models used by the software simulations against which the controllers are tested and validated.

Four major factors can impinge on the appropriate model fidelity selection level:

What question is being answered, or what problem is being solved? The strategic command level, for example, will tend to abstract away very low-level technical issues, such as the exact current position and velocity of a refueling tanker—while total and aggregate reports on friendly and enemy losses, and general information on force disposition, are often included in decision making. Of course, it is always possible that what might ordinarily be considered low-level detail may rise to a level of strategic importance, given the right set of circumstances—therefore, rapid, accurate access to such information must be available. Nevertheless, while for many routine decisions, the appropriate models can be highly aggregated and abstract, yet the closer one moves to the level of tactical engagements, the more important detail becomes, and hence the supporting models need more fidelity. Thus, the question of interest can influence the choice of model used.

What computational resources are available? As a general rule, the higher the modeling fidelity, the greater the associated computational cost. Thus, the size and speed of the machine on which the model and algorithm will be executed, coupled with the maximum latency (elapsed time) permitted to perform the computation, will impose constraints on modeling fidelity.

What development resources are available? Developing and validating high-fidelity models takes time and money. Often, such models are brittle, and small changes in a target system can result in major changes to the software model. Hence, organizations are rightly skeptical about undertaking large investments in models that will soon be out-of-date due to rapid technical innovation. While simply maintaining existing models can be a significant investment, models with less engineering fidelity can be rapidly assembled and easily modified or adapted to changing circumstances.

What kinds of computation and algorithm are intended? Here, professional control theorists' participation is critical, for they understand the computational and algorithmic consequences of decisions concerning modeling fidelity. Including a third-order term, for example, may preclude certain kinds of powerful analytic techniques that can be used only for quadratic representations. In such a case, a control theorist can help make the tradeoff between model fidelity on the one hand, and powerful analytic techniques on the other. Throughout this book, there are places where exactly this kind of tradeoff has been performed. In every case, the researcher has deliberately chosen a lower-fidelity model to cast the problem in powerful algorithms.

From the point of view of this book's research, the fourth point is the most interesting and relevant. Here, Figure 2-3 depicts the trade space facing the control designer. The high-fidelity models are to the right, including such characteristics as being computationally demanding (or impossible, for certain kinds of algorithms), brittle, expensive to develop and maintain, and possessing increased credibility (that is, the user is more likely to believe and rely on these models' results). To the left, characteristics include being computationally tractable (thereby enabling a greater variety of algorithms), adaptable, relatively inexpensive to develop and maintain, and possessing decreased credibility. The issue, then, concerns where to stand on this continuum—anywhere from the far right (high fidelity, but very limited algorithmic complexity), to the middle (a balance between the two), to the far left (highly sophisticated algorithms operating on relatively simplistic models).

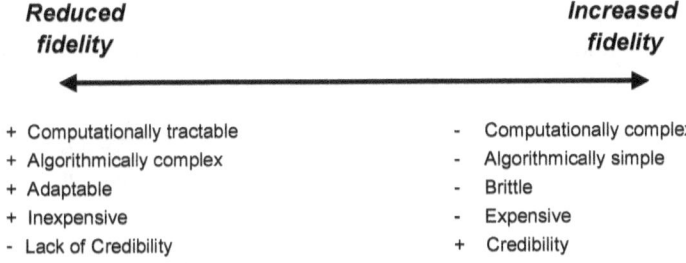

Reduced fidelity	Increased fidelity
+ Computationally tractable	- Computationally complex
+ Algorithmically complex	- Algorithmically simple
+ Adaptable	- Brittle
+ Inexpensive	- Expensive
- Lack of Credibility	+ Credibility

Figure 2-3. Modeling fidelity is depicted by the trade space facing the control designer, whose concern focuses on where to stand on this continuum—anywhere from the far right (high fidelity, but very limited algorithmic complexity), to the middle (a balance between the two), to the far left (highly sophisticated algorithms operating on relatively simplistic models).

Of course, it would be nice to have the best of both worlds, high-fidelity, engineering-quality models as the basis for mathematically complex and sophisticated algorithms. While such a desired result is not currently within technical reach, the point to remember is that, when one views dazzling, high-powered mathematics, there has often been a price paid in modeling simplicity. Similarly, when one sees elaborate, fully fleshed-out military campaign models, they may have been created at the cost of heuristics and severe algorithmic compromises.

MODEL PREDICTIVE CONTROL

The following discussion on Model Predictive Control (MPC) [1], addresses concepts of the rolling horizon, the difference between open and closed-loop algorithms, and, in this incarnation, includes implementation of stochastic optimization [2].

The Rolling Horizon

The key idea underlying MPC is that using a model to select a particular plant's control signal can result in predictive behavior over an appropriate time period. Generally, the signal that produces the best estimated result is selected for implementation.

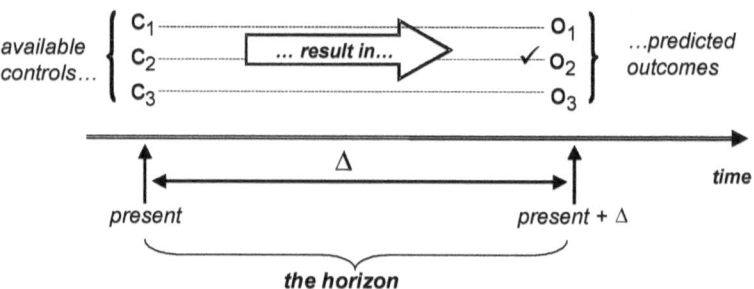

Figure 2-4. The key idea underlying MPC is that using a model to select a particular plant's control signal can result in predictive behavior over an appropriate time period.

As in Figure 2-4, let us suppose that three control signals—C_1, C_2, and C_3—are available for selection. Using a plant model and its dynamics, the MPC algorithm predicts what each likely outcome would be if each control were selected. The algorithm then evaluates the objective function against the predicted plant state and selects the control signal corresponding to the best outcome—outcome O_2, corresponding to control signal C_2—to issue to the plant.

A secondary factor, the horizon, is the prediction's length of time

(Δ). In some cases, a horizon's length can be considerably longer than we can wait before updating our control signal. Therefore, a key aspect of MPC is that it need not wait the entire horizon length before repeating the process. Such a situation is illustrated in Figure 2-5, which shows three successive, and partially overlapping, control decision operations.

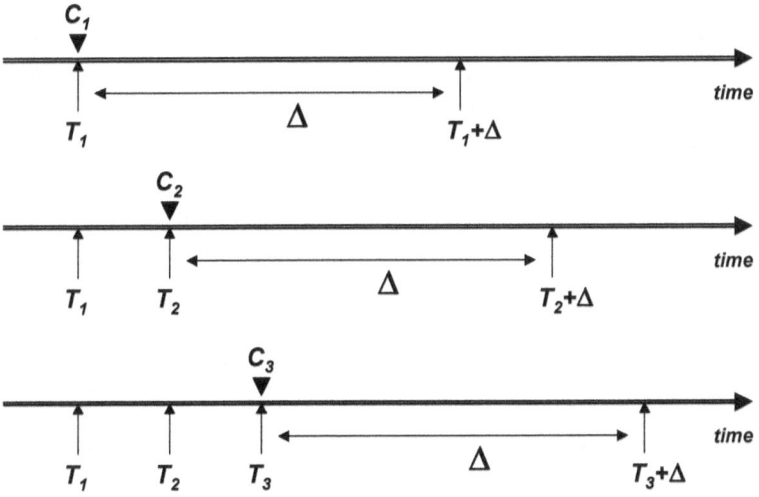

Figure 2-5. A key aspect of MPC is that it need not wait the entire horizon length before repeating the process; shown above are three successive, and partially overlapping, control decision operations.

At time T_1, the MPC algorithm executes, optimizing over the horizon [T_1, $T_1+\Delta$]; then control C_1 is selected for implementation. A short time later—short, that is, when compared with the horizon length Δ —the process is repeated at time T_2, and the control signal C_2 is selected by optimizing over the horizon [T_2, $T_2+\Delta$]. This process is then repeated again, at time T_3, as control signal C_3 is selected by optimizing over the horizon [T_3,

$T_3+\Delta$]. And so on. As Figure 2-5 indicates, three successive controls (C_1, C_2, and C_3) are issued at three successive times (T_1, T_2, and T_3), each as a result of an MPC optimization over a horizon of length Δ .

The situation depicted in Figure 2-5 is the classic example of the rolling horizon. As the algorithm reaches each successive new control point, the time horizon over which the MPC optimization is made rolls forward into the future.

It often is the case that the optimal answer for a short horizon is different from the optimal answer for a long horizon. But the longer the horizon, the more savvy the algorithm can be in trading short-term results for long-term gain. Yet, the longer the horizon, the more uncertain the predictive process becomes. Considering the application at hand, the horizon-length tradeoff is an engineering issue that will always be resolved on a case-by-case basis.

One difficulty for the algorithm developer is to ensure that progress is always being made toward the ultimate goal. If the horizon is too short, or if the objective function is poorly chosen, the algorithm can locally optimize but will not make progress toward the desired end-state. One way to deal with this problem is to choose a point in the future as the not-to-exceed time—that time by which the desired end-state should be reached. Then, at each control point, the length of the horizon will be adjusted to be long enough to cover this goal. Early on, the horizon will be long. But as time proceeds toward the deadline, the horizon naturally shortens. Throughout, the point is to ensure that the algorithm does not wander aimlessly trapped in a local minimum, but has enough forward vision to make progress toward the goal.

Open-Loop Versus Closed-Loop

In Figure 2-5, for time T_1 the MPC algorithm attempts to predict what will happen over the corresponding time horizon, [T_1,

$T_1+\Delta$]. Yet during the two subsequent time lines, the algorithm will make additional control decisions—that is, control points T_2 and T_3 lie within the T_1 time horizon allotted for its prediction. Thus, if the MPC prediction capability is to be accurate over its horizon, it must understand that additional control decisions will occur. In effect, the MPC predictive capability must be self-referential—it must know that it is part of the system, and that its decisions will affect future system behaviors. Put another way, a system with a predictive MPC capability will behave differently from a system without such a capability. Thus, to predict the overall system's behavior accurately, the predictor must know about its own presence and behavior in the system.

Not all rolling horizon predictors have such a self-referential characteristic. Instead, in doing their predictions they ignore intervening control points, modeling how the system might evolve if no controller were present. Such an MPC controller is called open-loop because its optimization models do not explicitly take into account that the controller itself is present in—and affecting the behavior of—the system. For example, the well-known military OODA model is an open-loop in just this way. On the other hand, if the predictive capability has within itself models that account for future control decision points, then the MPC has a closed-loop predictive capability. Most of the MPC-style controllers discussed in this book are the latter.

One great strength of the MPC approach is that in principle it is able to accommodate a very large variety of approaches to predicting system behavior. For any particular approach, there may be specific mathematical structures that can be algorithmically exploited; but the MPC idea, *per se*, does not depend on the presence of such structure. As one important example, it could be that the only plant model available is an event-driven computer simulation of the battlespace. But so long as the simulation is able to roll itself forward to estimate likely future events, that is all the MPC algorithm needs to function. Thus, MPC can be considered an encompassing architecture within which a large variety of specific modeling approaches finds a home. Such a situation is a great advantage because MPC can contain a large variety of available control and modeling approaches, providing a system architecture within which different approaches can peacefully co-exist.

Dealing with Uncertainty: Stochastic Optimization

As in Figure 2-5, MPC's natural mode is for the controller to be online, meaning that the algorithm is re-executed as necessary—and in real time—to generate new control signals. Here, for example, the MPC algorithm is executed once at time T_1, then—in real time—it is executed again at time T_2, and again at time T_3. Why does the algorithm need to be re-executed? Because events have presumably changed. While it took its best guess about what the future would be, it implicitly acknowledges that there will be some error in its approximation. Indeed, new information will have become available, so that a revised estimate using new data is appropriate.

Even if such a system were perfectly deterministic—so that, in principle, all controls could be generated in advance, and then implemented as a table—we might still choose to redo the computation as needed to prevent having to pre-compute the large number of possible controls. In military operations, however, uncertainty is inherent and unavoidable. Many effects—weather, engagement outcome, battle damage, enemy actions—are either inherently probabilistic or else well beyond any accurate model capacity. Thus it seems that an approach recognizing and modeling uncertainty in plant evolution has certain advantages over approaches assuming purely deterministic behavior.

Take, for example, a typical four-to six-hour period of time in an Air Operations Center (AOC). Events which could occur during this time horizon, and which might trigger modifications to the air tasking order, include aircraft loss, high-priority and/or time-critical target detection, weather changes, and unexpected enemy action and defensive capabilities. Therefore, it ought to be possible to construct probabilistic (that is, stochastic) models of their likelihood and incorporate such probabilities into initial planning.

Thus, given the right type of models, and taking into account underlying probability distributions and co-variances, it is possible to estimate future state probabilities, compute values for possible outcomes, and, based on expected events, optimize current control. This variation on the MPC theme adopts stochastic optimization. It is able to hold resources in reserve, betting on probabilistic models that the resources will be needed in the not-too-distant future to respond to events that have not yet occurred—the discovery of time-critical targets, for example. Of course, those events may not materialize; but if the probability model is correct, then more often than not the decision to withhold some assets will prove to be better over-all.

From a purely theoretical point of view, the algorithmic technique appropriate for this type of problem is dynamic programming (DP). However, unless the state space is highly simplified and abstracted— a technique with fidelity problems of its own—the computational complexity of DP for military operations problems is unmanageably large. As a result, DP solutions are generally relaxed. Rather than solving complex Bellman equations recursively backwards from the goal state (that is, the theoretical optimum), for example, algorithmists solve forward from the current state. Here, however, the algorithm will truncate the calculation with an estimate of the cost to go, for example, the likely result should this particular branch of the decision tree be taken.

The other approach to dealing with uncertainty is frequent state feedback. In this approach, one assumes, incorrectly, that the system evolves deterministically and then computes the optimal control. In addition, however, one also computes a behavior threshold, meaning that so long as the behavior of the system entities remains within the threshold, the computed control law is used as is. In effect, by remaining within the threshold, the system evolves as if it were deterministic to within the control law's assumptions. However, as soon as the system behavior crosses the threshold boundary, the system will compute a new, updated control signal to replace the previous one. Alternatively, another one more appropriate for the new circumstances can replace the current controller. If the non-determinism of the system is small enough so that threshold crossings are relatively rare, such an approach can succeed—even though it does not model uncertainty explicitly.

WHAT IS A GAME?

Here, we consider a most powerful theoretical tool: the ability to model the engagement as a game, in which two or more opposing players issue controls to their respective resources, attempting to maximize (or minimize) some objective; i.e., to win the game. Overall, game theory [3] provides a solution to a number of difficulties facing the control designer, while simultaneously introducing new problems of its own.

A major difficulty facing any researcher attempting to automate C^2 is how to deal with an active, intelligent adversary? Indeed, the existence of such an adversary is a limitless source of skeptics' critical material, since it appears that any control approach must first make a number of unrealistic assumptions about enemy behaviors—assumptions ripe for debunking by those with first-hand military operations experience. The adversary, it seems, is always full of surprises, and the very thought of accurately predicting adversarial actions, or of basing real military decisions on such predictions, appears ludicrous on its face.

However, the control designer can make two significant points. While the enemy is certainly unpredictable, not all of the enemy's actions are optimal, at least in terms of opposing our side's goals. Further, the game control theorist can formulate his own goal via objective functions—to increase (or maximize) our value while decreasing the enemy's.

The control designer can use game theory to state the worst the enemy can do, then the designer can create an adequate plan (or control law) to counter it. Such a concept of operation will necessarily help military operations face what the enemy actually does—since it will necessarily be less stringent than his worst. That is, the *worst* the enemy can do becomes a conservative lower bound.

Throughout, the control designer assumes that he and the enemy share an objective function: the control designer seeks to maximize his values, the enemy to minimize them. Working under that assumption, the control designer uses models of plant state and dynamics to devise two control laws, one for his own resources, one for the enemy's, both in equilibrium. That is, mathematically the designer can demonstrate that any deviation from his control law would be worse for his side, and any enemy deviation would be worse for him.

In reality the enemy may choose not to follow the control law the designer has computed for him—in fact, he may be totally unaware that the computation has occurred at all. However, the mathematics shows that any deviation by the enemy away from this optimized equilibrium will be to the designer's advantage. Because, once again, he's computed the worst the enemy can do—then formulated an optimal control strategy to counter it. Therefore, whatever actually happens in the battlespace will be better than his computed worst-case scenario.

Playing the Game

There are other advantages as well. First, by solving the game-theoretic problem, the researcher gets two control laws, one for friendly assets, another for the adversary. Even more, by treating both sides symmetrically, the game-theoretic approach does not play favorites. Instead, it forces both sides to play their best. Such concept, in turn, is a powerful rebuttal to the charge that control theoretics makes favorable assumptions about the adversary. On the contrary, by using game theory the designer has constructed a worthy opponent, one that plays the game as well as we do. Indeed, game theory eliminates the natural human tendency for wishful thinking. Once again, the designer systematically faces the worst-case scenario.

Constructing a control law for a worthy opponent has another advantage, enabling a simulation to play both sides. That is, rather than have a scripted or canned opponent forces (OpFor), the researcher devises an OpFor with a similar level of responsiveness and optimality to the control law proposed for the friendly forces. Such a usage allows the simulation to assess many different scenarios and battle conditions rapidly, confident that any generated enemy actions reflect and respond to changes in underlying plant characteristics. Indeed, finding a credible OpFor is the Achilles' Heel of wargaming, game theory offers a credible solution.

Returning to Figures 2-1 and 2-2, the game theory advantage is clearly illustrated. Imagine two researchers, each attempting to validate proposed control laws. The first, adopting an experimental testbed that looks like Figure 2-1, faces the immediate question: where is the adversary in the plant, and how have we modeled his actions? The best this theorist can say is that he has attempted to construct a few stressing cases. If his algorithm performs well, then there will be increased confidence that it will also perform well in the real world.

Compare this scenario to the second game-theoretic's argument using a configuration like Figure 2-2. Here, as part of the research itself, the theorist has constructed not only an optimized controller for the Blue side; he has simultaneously constructed an optimized controller for the Red forces. Rather than a static, scripted adversary, as in Figure 2-1, this adversary is agile and optimized—a worthy opponent that stresses and validates the advantages of a control theoretic approach. Indeed, the experiments' credibility is greatly increased in the second case—thereby answering major criticisms against control theoretic approaches.

Winning the Game

Another practical advantage of game theory is its consistency with the fundamental goal of developing formal models for plant state and dynamics. While it is not so hard to develop formal models for speed and lethality, it is indeed difficult to develop models for how an adversary will control his assets as the battle dynamically evolves. Currently, two widely used approaches are rules, to capture heuristics and doctrine, and reenactment of known adversarial actions from historical battles. However, neither of these approaches is formal—that is, neither is susceptible to analysis and optimization using the control theorist's mathematical resources. Instead, both have an *ad hoc* quality that inhibits generalization and deep analysis. Game theory, to the contrary, permits the designer to break through this heuristic barrier, developing a theoretical, mathematical model of the adversary's C^2 that is completely consistent with proposed approach for managing friendly C^2. In its lack of bias, such symmetry is powerful and intellectually appealing.

For practical significance, consider the concept of operations for a decision support tool shown in Figure 2-6. Here, the experimental framework suggested by Figure 2-2 has been augmented. Instead of interacting directly with the real plant—the battlespace—the friendly and adversarial controllers (computed and optimized using game theory) are playing the game against a simulated plant. It, in turn, is continually updated with state information from the real plant—the ongoing real military engagements. Thus, the simulated plant is a representation of the current state of the world, within which Blue and Red controllers play their game.

Because the simulated war is being played at computer speeds, a great number of variations can be played. Monte Carlo techniques can be used to generate credible probability distributions for key variables. Hypotheses about unobservable conditions can be proposed and verified—or disproved. Comparing simulated results against actual observations can test assumptions concerning key parameters—lethality, speed, and size of forces. In short, such a configuration becomes a real-time testbed for generating sensitivity analyses and hypothesis testing. What makes all this possible, and credible, is that we have a reasonable model of what the adversary might rationally do, over time, based on sound theoretical underpinnings.

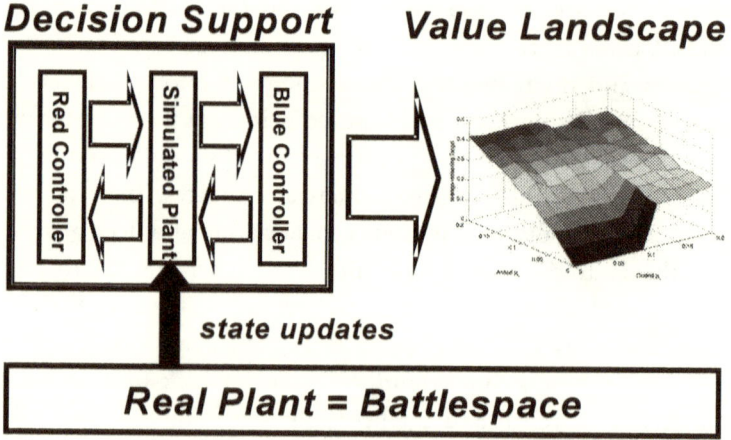

Figure 2-6. The friendly and adversarial controllers are interacting with a simulated plant that is continually updated with information from the real engagements.

Nash and Stackelberg Solutions

There are three major solution approaches to game-theoretic formulations, Nash, Stackelberg, and Abstract Board Games (ABG), such as chess.

The problem formulation key assumption is that both sides share the same objective function—one is trying to maximize this function by issuing appropriate control decisions to its assets, and the other side is trying to minimize the objective function by issuing control decisions to its assets. When both sides have the same objective function, it is a zero-sum game: whatever one side wins, by definition, the other loses. Since Blue knows its own objective function, it assumes that Red also shares the objective function—and is trying to drive it in the opposite direction.

A Nash solution, as in Figure 2-7, solves the zero-sum game. Here, neither side has any reason to change the solution, for any Red—or Blue—change will result in a worse situation.

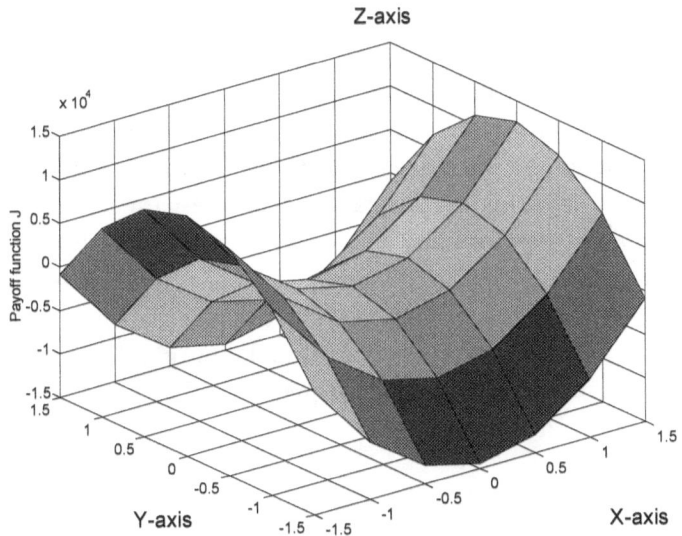

Figure 2-7. The Nash solution appears in the middle as the z-axis plots the game's value for various Blue and Red choices.

Figure 2-7 represents a simplified situation, in which Blue has only one control decision—to move back and forth in the x-direction.

Similarly, Red has only one control decision—movement along the y-axis. As the z-axis plots the game's value for various Blue and Red choices, the Nash solution appears in the middle. Blue likes it, because any deviation in the x-direction shifts the value of the game downward, opposite Blue's goals. Similarly, Red likes it, since any deviation in the y-direction shifts the value upward, opposite Red's goals. Here, then, a rationally computed control law represents the best possible outcomes—given an intelligent adversary.

In reality, both Blue and Red have many dimensions—or degrees of freedom—along which they can make control decisions. Despite a change in the problem's dimensionality, the fundamental idea remains the same: find control laws that represent a simultaneous optimizing of objectives.

When a Nash solution exists, it is both satisfying and of great theoretical and practical significance. However, not all games have Nash solutions, for the value functions' internal geometry does not correspond to the saddle-shaped template. If that is the case, the control theorist imagines a series of alternating decisions— first by Red, then by Blue, and so on—supposing as before that each player has an objective function which he is trying to optimize. Perhaps it is the same objective function—a zero-sum game— perhaps not. In any event, the Stackelberg solution under consideration operates even in the presence of small differences— almost zero-sum games.

In this model, for each move, the player selects a control decision that optimizes his objective function. However, since the player knows his opponent's objective function, one factor he takes into account is how the opponent can be expected to respond. That is, by selecting a control law, and an action that tends to push Red in a desirable direction, Blue influences Red's behavior. If Blue finds such a course of action (COA) desirable, Red must also feel it is desirable—since Red actually makes the decision. Blue, in effect, attempts to arrange things so that, for Red, the least of all evils is an action Blue favors. And so on. The alternating Blue-Red-Blue decision-making process, called leader-follower, has Blue attempting to lead Red by Blue's own control decisions.

There is considerable technical literature comparing Stackelberg and Nash solutions. While Nash solutions' great advantage is an associated mathematical structure, Stackelberg solutions, although less susceptible to rigorous analysis, are always computable. Both approaches have been applied to military operations, often with remarkable improvements over controllers developed without using game-theoretic techniques.

Adversarial Intent: The Hidden Variable

Adversarial intent, a most exciting recent research topic, is, in its abstract mathematical formulation, a special parameter estimation. Using plant identification, state initialization and update, and adaptive control, game theory enables the controller to engage in a kind of psychoanalysis of the adversary and his intent.

Here, formal model mathematical entities are in two classes, depending upon the rate of change over the life of the controller. Some variables, called parameters, vary not at all—or at least very slowly—over a controller's useful life. In effect, they can be treated as constants that, set in advance, never again need be altered during subsequent calculation. Rapidly changing variables, called *state variables*, require that the controller track current changes, predict future changes, and then make control decisions that will drive key variables correctly. State estimations, the value estimation of state variables performed at particular points in time, are a concern. For example, an aircraft's position and velocity is a state variable, which changes rapidly over time, while an aircraft's maximum speed, and its maximum fuel capacity, are parameters, and are generally treated as non-time-varying constants.

Plant identification, what the control community knows as selecting model type and setting associated parameters, involves measuring the behavior of the system to be controlled, and of its constituent entities, and estimating parameter values (using, for example, a least-squares model matching approach). For example, to find a particular aircraft's maximum speed, one might look in open literature on weapons systems, then record the value in a table (or even embed it directly in software code). Some quantities, however, are not directly observable, and can only be deduced indirectly from other variables. An example might be the amount of jet fuel at a hostile airbase. By observing the rate at which fuel trucks enter and leave the base, and the flight patterns of the base's aircraft, a reasonable fuel availability estimate could be deduced, even though the enemy is unlikely to allow observers access to enclosed storage tanks.

Further, even though we may assume that a parameter's value is constant, that assumption is inexact for two reasons. First, even if the parameter's true value does not vary, our knowledge about that parameter—our best estimate of its value—can change as the plant evolves over time and we collect new information. Second, some quantities treated as parameters do, in fact, change over time. One example is ground-troop lethality: as they gain experience, troops get better, becoming more effective adversaries. If lethality is not updated, modeling accuracy will correspondingly degrade.

Adaptive controllers, those that explicitly recognize the need to update parameters and state variables, do not treat plant identification as a one-time process during design and initialization. Instead, adaptive controllers treat parameter values adjustment as a continuing process—almost as if plant identification were continually executing as a background process.

Throughout, game theory employs an adversarial model in an objective function. In a zero-sum model, the objective function is assumed to be the exact mathematical negative of one's own intent.

Yet in all likelihood the adversary's actual intent is not *exactly* our opposite. In fact, if we were to allow the adversary to set his own objective function parameter values, his numbers might vary—in small ways, certainly, but perhaps in large ways as well—from values we would assign based on our objectives.

The question then becomes what *is* the enemy's objective function? If we assume that our adversary is acting rationally to achieve certain well defined objectives. And if we further assume that those objectives can be reasonably captured by an objective function. Then it follows that, by observing actual command decisions our adversary makes over time, we should be able to estimate his objective function parameter values—parameters not directly observable which the adversary would almost certainly prefer we not see! Indeed, the mathematics involved in such calculations are not different, in kind, from state estimation calculations (for time-varying state variables) or adaptive control (for slowly varying parameters). Models of plant state and dynamics causally link observable quantities to hidden—that is, not directly observable—quantities, and least-squares estimation approaches (such as Kalman Filters) can be used to take out noise and reduce uncertainty.

PEERING INTO THE FUTURE

Once we possess a reasonably good approximation of the enemy's objective function, we are in a much better position to predict future behavior. Further, as soon as enemy actions begin to deviate significantly from predicted values, we are immediately alerted to a shift in intent, and are then ready to speculate about the enemy's sudden change.

It may be asked whether or not the enemy could perform the same kind of analysis of our actions. While the theoretical answer is yes, the practical answer may be somewhat different depending on computational resources. Although the kind of estimation we've been discussing is extremely computationally demanding, the

United States, as the world leader in supercomputing technology, enjoys a unique position in its ability to carry out all required calculations.

To test such ideas, one would perform a computational experiment using a set-up much like that in Figure 2-2, but with a simulated software plant. With different objective functions introduced into the Red and Blue controllers, and using a zero-sum assumption about the opponent, each then does a full game-theoretic optimization. Beginning with both sides issuing controls into the simulated plant, Blue observes the state as it evolves and attempts to infer Red's true objective function. In toto, this is an elegant, well defined, and potentially revolutionizing application of game-theoretic optimized control.

Nevertheless, it is a subtle matter to determine what use one should make of the information produced by such calculations. Indeed, it would almost certainly be a mistake to feed such estimates directly back into the controller, for to do so would provide the enemy with an opportunity to lull us into a false sense of security. Rather, this technology's appropriate concept of operations should be hypothesis testing and verification. The algorithm can compare predicted enemy behaviors against observed behaviors, detect significant discrepancies, and suggest possible explanations. Such analyses should then be combined with other information sources as part of a general intelligence analysis process that also seeks independent confirming data. Thus, this type of algorithmic information is only one piece of a larger puzzle and should not be used blindly or in isolation. Nevertheless, formal adversary modeling *does* enable this sort of close analysis, thereby providing significant value-added to the intelligence process.

LINEAR MODELS IN A NONLINEAR WORLD

While MPC and game theory techniques are important, and employed by many researchers, by no means do they exhaust the breadth of available technical approaches. Following, we discuss four alternative modeling and optimizing methods, including using traditional linear techniques on this highly nonlinear military problem, abstract board games to model tactical engagements, discrete event systems, and emergent behavior.

Using linear models [4] in control theory has many advantages, including the fact that the developed theory for linear systems is extensive, as are commercially supported algorithms and software toolboxes. Thus, in using linear models there is a great incentive to sacrifice modeling fidelity somewhat, while building as good a linear model as one can, then hoping that frequent state updates via the closed loop will suffice to keep overall behavior reasonably satisfactory.

Integer programming, a variant of linear modeling used extensively in the commercial world to solve resource allocation problems (crew scheduling in commercial aviation, or route planning for parcel delivery), can tackle large problems in reasonable time frames. Indeed, since the construction of any military tasking order is in essence a resource allocation problem, an IP approach could be highly successful.

Briefly summarized, a particular resource residing at a particular place and time may be represented as a 1 in the cell of a very large array listing all possible combinations of entities, times, and places. A zero (0) in the cell represents absence. Next, constraints are formulated to ensure mutual exclusion and route feasibility, while an objective function—linear in the key variables—measures the goodness of any particular feasible solution. Since the answers in the cells are constrained to take on the discrete values 1 or 0, ordinary continuous liner programming techniques do not work. Hence, the program requires a large tree search over various combinations of cell entries, with each node requiring its own solution to an associated linear programming problem, itself solved using a standard linear programming (for example, Simplex) algorithm.

Although it would seem that the above is a reasonable approach to solving the problem, a recent attempt at this type of implementation proved frustrating. Research began by constructing a complete IP formulation of the problem adequate to produce an exact solution. However, even after valiant efforts at pruning and optimizing the algorithm, this problem's run time was excessively high. Then programmers began a series of substitutions, exchanging computable heuristics for aspects of the non-computable exact solution. In such an undertaking, because there is no formulaic method available, engineering judgment and experimentation suggest and verify particular approaches to particular bottlenecks. The result is an IP formulation that contains within it a fair number of heuristic approximations, demonstrating the engineering tradeoffs that arise when linear techniques are applied to a nonlinear problem.

Another research team used another variant involving a Lanchester-style attrition model for troop engagements. To linearize the problem, the engagement area was hexagonally gridded, and an objective function formulated which would lead Red and Blue forces to traverse the field, reaching and attacking the adversary's fixed assets. The result was a sequence of moves on the gridded field in which all forces would avoid and/or engage one another optimally so as to achieve the objective.

The difficulty was that the linear model would occasionally produce infeasible results—as, for example, when total available resources in a given cell were computed to be negative. Setting such threshold boundary conditions, while easy to write, has the unwanted side effect of altering the mathematics so as to require an entirely different—and far less tractable—algorithmic approach. While much of the time the algorithm produces excellent results, occasionally such infeasible results appear.

The answer turned out to be creating surrogate metrics, with ability to achieve the desired behavior while being compatible with the optimizing linear approach. For example, rather than use a count of friendly and enemy targets killed as part of the objective function, the distance between the units could be used instead. The operative idea is that closer units will engage, hence attrition will occur. Indeed, if we use distance squared as the surrogate measure, the solution's non-negativity is maintained without sacrificing the objective function's convexity. This and similar devices enabled the researcher to formulate an objective function that produced reasonable behavior but was also amenable to linear solution techniques.

Along these lines, the linear versus nonlinear tradeoff will continue to be a major area of applied research—in part because the power of linear techniques is simply too great to ignore. Yet the resulting linear approximations can, without additional clever engineering, produce unacceptably infeasible or computationally extravagant results.

BOARD GAMES AS PARADIGM

Just as game theory can be used to construct optimized controls based on the opposed minimization/maximization of selected objective functions, so too, in addition to the more numerical approaches, there is one similar to chess-playing algorithms which models an engagement as moveable pieces on a game board. Here, for example, velocity constraints are modeled by the relationships between the cells' size and the distance a piece can traverse in a given move. According to deterministic or probabilistic game rules, interactions between pieces result in attrition, while winners and losers are determined by the end-state achieved.

Although many models use gridded areas and discrete time (or moves), the unique aspect of an ABG is that the control algorithm plays out the game several moves ahead, keeping track of attractive play lines, then selects the next move (or control signal).

Standard ABG techniques rely on variations of exhaustive searches, where *all* possible future legal move sequences are examined in advance. While the number of future moves, called the *depth* of the search, may seem small, it is not. For even a small number of pieces on a fairly small board (such as chess), the resulting search tree becomes prohibitively enormous after only a few moves. Currently, however, even the best of the world champion chess-playing programs can look only 10 or 12 moves ahead. Therefore, for a problem as complex as a military engagement, with the game board size approaching several thousand cells, such an exhaustive search is simply not a realistic algorithmic option.

Linguistic geometry [5], one alternative algorithmic approach, relies not on a brute-force exhaustive search but on a constructive search, where heuristics are used to suggest the most promising paths for consideration. An opponent's moves to block, or divert attention from, such paths are then computed, followed by possible counter-moves. By using such a constructive approach, considering in detail a few promising scenarios, the algorithm is able to suggest and develop good solutions while greatly reducing computation.

It is said that the game of chess was created as an analogy of wars fought before modern weaponry. Since algorithmic techniques were developed out of detailed studies of chess, there is a certain intellectual satisfaction that chess may have brought us full circle, providing practical solutions to modern military operations' C^2 problems.

DISCRETE EVENT SYSTEMS (DES) AND HYBRID CONTROL

There are other, very different, very powerful, ways of thinking about such a system and how to control its behavior. First, we do not approach the entire plant (the whole battlespace), but rather consider individual entities with key dynamical behaviors. For an air operations problem, for example, such entities would include aircraft, detection radars, SAMs, and so on. Next, we broaden our notion of state to include not only measurable variables (such as position, velocity, and fuel remaining), but also the entity's condition at a given point in time. Further, state might include such conditions as: at-base, in-transit-to-target, detected-by-hostile-radar, authorized-to-release-weapons, damaged, and so on.

Therefore, if we look at a single entity over time, we see a sequence of transitions from condition to condition (or state to state) describing both the events that took place affecting the entity and the actions it took in response. In short, the sequence would read like a time-tagged event log, marking both the points at which transitions occurred and the new state resulting from each transition.

One way of representing this thinking is a graph, where the vertices represent possible states (or conditions) in which the system might find itself; and the edges (directed edges, with arrows) represent allowed state transitions (from one state—detected-by-enemy-radar—to another—under-attack). Both discrete event system and finite state machine (FSM) describe this way of posing and representing the system model [6].

Now some state transitions are not under our direct control; instead, they simply reflect an external event to the modeled entity. Other state transitions, however, result from decisions (or controls) made at or by the entity. For example, an aircraft controller decides when and whether to transition from at-base to in-transit-to-target. In this modeling, then, the controller's major task is to decide—among various state transitions available now and in the future—which to select. Guiding such a decision will be both the desire to achieve useful goals (enemy-target-destroyed) as well as to avoid undesirable consequences (aircraft-out-of-fuel).

Part of the controller's design includes constructing the state model to describe the system (that is, selecting named vertices and allowed edge transitions). Clearly, such a model depends heavily on the entity's characteristics as well as the types of task it will be asked to perform. However, the great advantage of modeling the system this way is that it attempts to create a complete enumeration of all possible states—and state transitions—in a form susceptible to automatic processing.

Carrying this idea one step further, it then becomes possible to map discrete event system representations at low levels (tactical entities, say) onto higher command levels (commanders, say, or an operations center). Here, the term map indicates that states and state transitions at the lower, detailed level are aggregated or abstracted into states and state transitions at the higher level. Such an aggregation also has the benefit of shielding upper levels from unwanted detail, and for purposes of operational planning, treating many individual tactical entities as a single entity. Along these lines, one interesting recent research area is an attempt to automate the construction of such maps through an information theoretic analysis of actual event logs generated by live entities under real battle conditions. Such event logs contain a wealth of information about entity behaviors, about what is and is not most important in both triggering state transitions and appropriately aggregating data. Being able to automate what is, at present, a manual process would greatly accelerate the rate at which effective DES models and systems can be constructed.

The use of DES techniques is especially appropriate when the tactical entities are autonomous or robotic.

It is possible to combine traditional continuous optimized controllers with DES techniques to create what are called hybrid controllers. Here, a different control law (or optimized online controller) is associated with each of the DES states. When the system is in a particular state, the associated controller is activated;

however, when the system transitions to a new, different state, a new controller replaces the previous one. Such a set-up permits a single system to apply a wide variety of control techniques, each optimized to the system's current conditions. For example, a value landscape analysis can show that a stochastic optimization is appropriate in one circumstance (when the enemy is largely passive, say), but that a game-theoretic controller is appropriate under other conditions (when the adversary is present in force).

In addition, the DES can impose higher level reasoning to detect such differences, and can then select the right controller for the case at hand. Such an approach provides a powerful merging of artificial intelligence techniques with those of traditional control theory. Indeed, it seems clear that future deployed control systems will exhibit exactly this kind of hybrid architecture.

BIOLOGY AND EMERGENT BEHAVIOR

Finally, some techniques for the control of military operations also use complexity theories based on biological analogies [7].

As an example, consider the way ants are able to locate food sources and then communicate to colony members the best path to take to the food and back again. When a wandering ant finds food, it begins to deposit a highly specific and recognizable substance called a pheromone.

In the meantime, ants near the colony are depositing a different but also recognizable pheromone associated with the nest. The ant with the food then begins to follow the nest pheromone back, while ants without food begin to follow the food pheromone toward the newly discovered source. At first, the food signal is weak, but as time goes on, and more ants participate in the process, the pheromone trail increases in intensity. Further, it need not be a straight line. If there are obstacles in the way, the path will naturally avoid them while still solving a minimal spanning tree problem.

Such a phenomenon, which occurs repeatedly throughout the natural world, is called emergent behavior.

Because pheromones dissipate into the ambient atmosphere, when a food source is exhausted, the path to it disappears. Thus, the mechanism naturally forgets as information becomes out of date. In an adaptive, optimized manner, new food sources—and associated pheromone paths—replace old ones. Clearly, the colony is acting in a highly intelligent and efficient manner, even though any individual entity is blindly obeying simple local rules of behavior.

A way to exploit these ideas is to associate different types of pheromones with such entities as enemy threats (aircraft, SAMs) and targets. Friendly entities (whether real or simulated) are attracted to the targets and repelled from the threats. The resulting emergent behavior solves the route-planning problem the same way the ants locate food.

Still, a number of technical and practical issues remain to be resolved, with perhaps the most novel solution being the use of place agents. Place agents may be small, inexpensive sensors (perhaps the form factor of a soft-drink can) scattered across the area of interest. Each would be, in effect, a bookkeeper for local pheromone information, aggregating and dissipating both threat and target pheromones over time, communicating such information to other place agents and nearby friendly assets.

From a military point of view, such an approach to solving the route planning problem has a number of advantages. It is very robust to the occurrence of conflicting or missing data, since it both remembers and forgets as the pheromones disperse across the field of engagement. It is very secure, since the global solution never explicitly resides in any one place, but is only implicit in the pheromone gradients spread across the aggregate of all place agents. It adapts rapidly and smoothly to new information (the unexpected detection of new threats; the disappearance or destruction of others). And it emerges out of a context that does not have to be modeled in advance—a very real operational difficulty associated with many techniques discussed.

There is a detailed working example later in the book.

SUMMARY

Control theory is a key field of study applicable to technologies for C^2. Applicable concepts include plant—the process and entities of military operations, and controller—the entity that accepts observations, makes decisions, and issues signals. Controllers require a formal model of the plant in order to make decisions. In such models, a snapshot of the plant, expressed as a collection of the plant's parameters, is called a state. The laws of how the plant evolves from one state into another are called plant dynamics. It is possible to model a plant at different levels of details and fidelity: greater fidelity allows more accurate predictions, but demands more computations.

To select among multiple alternative decisions, controllers perform a form of optimization, using models to predict the outcome of a decision. An objective function, also called cost function, is used to evaluate the goodness of an outcome. Objective functions often involve weights—relative importance of various aspects of an outcome. Constraints—formal expressions of what is feasible or allowed—also participate in evaluation.

A common technique—Model-Predictive Control—continuously adjusts control signals by predicting their effect within a rolling-time horizon. Many classical control methods rely on linear models that often are not good approximations for the nonlinear world of military operations. Other control approaches model the world as DES, or as Hybrid Systems, combining both discrete and continuous views.

Unlike traditional applications of control theory, military applications must account for an intelligent adversary and other sources of uncertainty. Game theory is a field of study that devises rigorous solutions for predicting optimal controls in presence of such adversary. Stochastic optimization is used to make optimal decisions in uncertain situations that contain a degree of randomness.

CHAPTER 3

PATHOLOGIES IN CONTROL: HOW C² SYSTEMS CAN GO WRONG

The previous chapter presented some key control theory concepts, approaches, and techniques that might be applied to C^2 systems. This chapter's central topic is, what can go wrong with C^2 systems that control theory could help fix? To frame an answer, the chapter will first examine a number of undesirable behaviors that occur in C^2 systems, demonstrating analogies between such pathological behaviors and phenomena studied in control theory and related disciplines. Second, the chapter will describe how designers of conventional control systems predict and mitigate such phenomena. Both presentations will be fruitful, for although C^2 systems exhibit many profound complexities and challenges not found in most conventional control systems, there are indeed numerous ways that a control theoretic perspective will aid both the study and design of C^2 systems.

Military analysts often attribute failures in C^2 systems to human error [1]. In many cases, however, C^2 failures have less to do with human error than with the inherent complexities of large-scale dynamic systems. Because failures in complex systems are themselves the subject of major studies in control theory and related disciplines, such as DES and computer science, it is reasonable to assume that these other disciplines can offer insights into—and solutions for—problems in C^2 systems. After all, damage mitigation is precisely what a well-designed control system is intended to do—to take corrective action when errors and disturbances happen—before they produce unacceptable ramifications or grow to unacceptable proportions.

It is tempting to devise mechanisms for preventing problems in C² systems (or any system, for that matter) on a case-by-case basis, rather than to derive solutions from a more general theoretical perspective. Such an ad-hoc approach might work for simple situations, including the following short and simple examples. However, it is difficult to accept the claim that all the problems that can arise in the design of complex C² systems should be resolved merely by applying simple common sense. Even conventional control systems, relatively simple and relatively easy to understand, produce complex and remarkably counterintuitive behaviors that require sophisticated analysis and synthesis techniques. Therefore, it is only logical to presume that there will be more of this type of behavior for C² systems that, by definition, are large, complex, involve ill-understood human dimensions, and an intelligent adversary.

Like control systems for large-scale technical processes, decision making in C² systems is hierarchical, and this structure is reflected in many of the examples in this chapter. Using the abbreviations DM for Decision-Maker, HLDM for Higher-Level Decision-Maker, and LLDM for Lower-Level Decision-Maker, the assumption is that LLDMs are subordinates of an HLDM. In the following examples, LLDMs are presumed to command a unit of military force that exerts direct physical impact, positive or negative, on other units, friendly or enemy. There are no specific assumptions about the service, scale, or type of units. They can range from an individual platform to a large collection of assets, from sea to air to ground, from reconnaissance to close combat.

In the following examples, there is no assumption that the C² systems necessarily involve computerization, automation, or high-technology communications. Since the presumption is that errors and other pathological behaviors are endemic to C^2 systems, technology per se does not bear on the basic thesis. Here, for example, the C^2 system could indeed use purely human decision making and venerable means of information gathering and communications. While it is presumed that the ongoing introduction of sophisticated technologies into C^2 will only increase the likelihood that pathological behaviors will occur more frequently, such pathologies are inherent in the nature of C^2— whether operated via smoke signals or supercomputers. Nevertheless, it is certain that with the ever-increasing role of technology in C^2, there will be a concomitant need for increasingly rigorous approaches for understanding, predicting, and avoiding pathological phenomena in C^2 systems.

TIME DELAYS

For many years, time delays have been studied as one of the classic problems in feedback control theory. Long before the advent of computer control, so-called transport delays—or, those delays in sensing the effects of control actions—had to be mitigated in process control systems. Typically, transport delays occur when sensors are downstream from the point of control influence, thereby introducing a delay in the information received by the controller. For example, such delays are experienced every time one takes a shower: the effect of a change in the hot or cold faucet position (a control action) isn't felt until the water flows through the pipe, out the showerhead, and to the skin (the sensor). In this case, if the delay is not taken into account, the target temperature will be overshot and undershot several times, making water-temperature adjustment a tricky and uncomfortable procedure. In C^2 systems, delays typically occur in the communication channel—and the effect is the same: the commander does not know immediately the effects of decisions.

Time delays are a key source of instabilities—often frightening and counterintuitive system behaviors that quickly diverge from normal, sometimes with violent oscillations. C² systems must contend with many sources of time delays, including the time required to:

— Collect information in the battlespace
— Assess and aggregate the information into a form suitable for presentation to an HLDM
— Transmit the information through the layers of DMs
— Process and evaluate the information
— Collect additional data and verify all the information
— Perform and coordinate all decisions
— Issue orders to LLDMs
— Make the necessary decisions and preparation processes at the LLDM level.

All these factors add to the significant delay between the time a situation is observed in the battlespace and when the control action is executed. Indeed, by the time the action is executed, the situation may have changed, resulting in such possibilities as missing an opportunity to gain an advantage, not countering threats on time, or not lending timely support to friendly troops.

In all ages, military practitioners have attempted to reduce time delays—by organizational mechanisms, training, and technology. However, overall time delays are probably not shorter now than they were in the days of Caesar or Napoléon, in spite of infinitely improved means of communications. The advantages of faster communication can be more than offset by increased complexity and demands in other aspects of the overall command chain. A number of technologies discussed in this book strive to minimize such delays. Chapters 6, 9, and 10, for example, look at various ways to enable shorter-cycle, continuous dynamic military action replanning and rescheduling, while Chapter 5 proposes a way to reduce delays through the radical decentralization of decision making.

Traditional feedback control systems deal with time delays in two principal ways. One approach is simply to de-tune the controller; that is, making the control action far less aggressive. Such an approach makes it possible to assess the effects of control actions before the actions go too far in compensating for perceived errors in the controlled variables. The second approach is to base the control action on a prediction of its effect rather than to wait for information from the sensors. In this approach, the signal used to evaluate the controller's effect is actually the output of a model inside the controller that gives an estimate of the effect immediately. This way, the process can be controlled as if there were no delay, and the model predictions are adjusted appropriately by actual sensed values. Returning to the shower analogy, both of these approaches are familiar. When in an unfamiliar setting, such as a hotel, one makes small adjustments to bring the temperature slowly to the desired level. This is a cautious de-tuned control strategy. However, in a familiar shower, one makes more bold adjustments before the effect is actually sensed because one has a good predictive model of what the effects will be when the faucets are turned a certain amount.

In C^2, using a de-tuned approach to deal with time delays—that is, taking small, incremental actions until the desired result is achieved—is usually not viable. An intelligent enemy will find a way to recognize and exploit such cautious actions. The real potential for enemy counter-measures is one of the reasons that so many of the technologies proposed in this book pay such particular attention to predicting possible enemy responses. In control theory, predictive models to deal with time delays are called Smith predictors, after the man who captured the concept mathematically. For C^2 applications, models—both mental and computerized—of friendly and enemy forces can be used to predict the effects of decisions. To that end, Chapters 4, 12, 13, and 14 discuss models capable of predicting enemy actions, while Chapter 8 pays particular attention to predicting the outcome of an engagement.

Of course, decision aggressiveness must be balanced against trust: how much credence should one put in these models? When there is uncertainty about how well models can predict effects, commands may have to be de-tuned; that is, caution is used so that real data can be received before too many commitments are made. Yet, the degree of caution must be balanced against the concern that it will provide the enemy with additional time to deduce friendly intent and take counteractions.

SAMPLING RATES

Sampling rates are related to time delays. Typically, in a C² system, as well as in many conventional systems, information about the state of the battlespace is provided to the HLDM as a time-sample—a snapshot—of the environment at regular, or approximately regular, intervals. The HLDM then interpolates between these samples to infer the state of the environment at all times. Since the state of the environment may actually vary widely between sample times, the HLDM's decisions may result in ineffective LLDM actions. Therefore, an HLDM's natural reaction would be to seek faster sampling rates. However, such a course is not necessarily beneficial or possible, for in most cases excessive sampling rates require greater efforts to collect and process information, leaving fewer resources to deal with other demands.

When information is available about process behavior only at sampling times, something must be done to assure that behavior between samplings is acceptable. In addition, any decisions or actions between samples need to be based on a reliable model of what the process is doing during the intervals when no fresh information is available. If possible, the chosen sampling rate will be fast relative to the process dynamics, so that the signal varies only slightly from sample to sample. Increasing sampling and replanning frequency—without causing thrashing—is one of the objectives of the technologies presented in Chapters 6 and 9.

If the sampling rate is slow relative to the process dynamics, there could be large changes in signal values between samples. Indeed, signals might even go up and down between sampling times, leaving such variations completely undetected. To avoid such undesired behaviors that cannot be detected in the sampled signal values, control engineers attempt to develop accurate models of the process, thoroughly analyzing the models off-line to make sure the sampling rate is sufficiently fast. Indeed, such an analysis must be performed before putting any control system online; that is, into actual operation. In control theory, such an analysis is based on an evaluation of the time constants in the process dynamics.

In C^2 systems, decisions are made to commit limited assets and with significant risks. Therefore, when the sampling rate cannot be made fast enough, predictive models can help, both to assess the current situation correctly and to gauge the effects of C^2 decisions.

CONTROLLER GAINS

The fundamental problem in feedback control theory is selecting the gains in the controller. In the simplest controller, a single controller gain is applied to the system error (the deviation of the system response from the desired response), defining the control action that is supposed to drive the system error to zero. Selecting the magnitude of the controller gain is a classic design tradeoff. On the one hand, a higher gain usually means a faster, more aggressive response to eliminate quickly undesirable deviations in the variables being controlled. On the other hand, a high gain can cause the system to overshoot the target, making it necessary to take corrective action in the other direction. When the controller is too aggressive (that is, the gain is too high), the system either becomes marginally stable (such as, it goes into sustained, undampened oscillations), or it even becomes unstable (for example, the oscillations begin to grow). High gains can also lead to control commands that exceed the control hardware's capabilities, so limits are hit and signals saturate.

In a C² system, a high gain problem manifests itself when the HLDM's orders cause an LLDM's actions to exceed desired limits. Such a problem is especially prevalent when the system is forced to operate close to its limits. (In military systems, it is almost always the case that one or both opponents operate near the limits of their capabilities.) The troubling results of such action could include placing the commander's own assets in an untenable situation, endangering other units, causing political complications, expending resources that could be better used elsewhere, and so on. In such cases, military historians speak of lack of caution, insufficient planning or intelligence gathering, misinterpretation of intelligence, poor judgment or arrogance, under-estimating the enemy, and so on [2]. Regardless of the perspective of military history, control-theoretic formalism would say these problems are caused by a gain being too high. Further, whether it is a purely human command or computerized system, the concepts and analytical techniques related to such high-gain issues can be useful in analysis and design of C² systems.

Similarly, an overly cautious or hesitant commander is analogous to a controller with low gain. Here, the results of low-gain control in a military environment could be a failure to exploit a time-critical opportunity, an inadequate response to a threat, the tardy achievement of a desired objective, or allowing the adversary to gain an advantageous position. Once again, the analytical and design techniques developed for avoidance of low-gain problems can be applied for the avoidance of similar problems in C², especially in those cases when a C² system involves a degree of automation.

For dynamic processes that can be adequately approximated by linear models, control theory offers many classic techniques for choosing gains in the feedback loop, thereby obtaining good balance between achieving favorable response speed and avoiding undesirable overshoot, oscillations, and instabilities. More recently, control theorists have developed methods for dealing with system limits and saturation, making it possible—in the presence of such nonlinearities—to tune the feedback gains optimally. For more complex systems, extensive simulation studies are used to evaluate and tune the gains in the feedback loop. There are also so-called robust control techniques that make it possible to select the controller gains, so that system performance will be acceptable for specified ranges of parameter values in the system model as well as external disturbances. A key source of nonlinearity in models of C^2 systems is the enemy, and for this reason Chapters 12 and 13 deal with formulating proper gains that take into account possible enemy responses.

Such off-line methods for control system design require adequate models of process behavior. For C^2 applications, where such models may not be readily available, online methods for tuning feedback gains may be of more value. Here, control theory provides some direction, if not complete algorithms that can be applied immediately to C^2 problems. As such, there are two basic approaches to adjusting feedback gains online, both of them falling in the general domain of adaptive control. One approach, called direct adaptive control, depends on a direct connection between observations and adjustments to the feedback gains. Using observations of the effects of past control actions to modify the feedback gain, these modifications might be stated as rules, such as: "When oscillations are observed for a specified duration, reduce the gains by a specified amount until the oscillation is eliminated."

The second approach, called indirect adaptive control, uses observations to update a model of the system being controlled. Then, based on this model, and typically using a method of off-line design, feedback gains are computed. Here, as in direct adaptive control, adjustments to feedback gains may be made only when some undesirable behavior is observed to avoid unnecessary computations and oscillations in the controller parameters. Several of the technologies described in this book implement adaptive strategies. In particular, Chapter 5 argues for a highly distributed approach to adaptivity, while Chapters 12 and 13 describe controllers that dynamically and automatically adjust their models of adversary as they execute their fights. Along those lines, Chapter 4 demonstrates how continuous adjustments in the model of enemy behavior can be used in a decision-support mode.

POSITIVE FEEDBACK

In control theory, the term positive feedback covers a broad class of instabilities in which errors get magnified rather than attenuated in the feedback loop. Or, to put it somewhat differently, the controller's corrective action is applied in the wrong direction, making things worse rather than better. Here, the system undergoes a self-reinforcing cycle of deterioration. Sometimes such a cycle is caused by a gain in the feedback loop that has the opposite sign from what was assumed in the system design. As a simple but common example, positive feedback can occur in industrial control systems when the polarity of a connection is reversed by inadvertently switching the wires for a voltage or current signal. Consequently, control actions are in the wrong direction—and, as errors are made larger rather than smaller by the feedback loop, the system becomes unstable. This example is in contrast to the instabilities described in the previous section, where it is the magnitude, rather than the direction, of the control action that causes instability.

A subtler source of positive feedback occurs when a feedback loop develops that was not anticipated or intended in the system design. One example of an unintended—and detrimental—feedback loop is the familiar squeal of an auditorium's sound system. In this case, the microphone picks up sound from the speaker, which in turn painfully reamplifies the stray sound.

As a C² example, suppose the HLDM issues orders to reinforce his forces in a certain counter-insurgency operation. Next, increased assets in the area require additional logistics and support installations. Yet, such additional support infrastructure creates more attractive targets to the insurgents. So, to deal with the increased attacks on the logistics installations, the HLDM is compelled to add more assets. And the spiral continues, as the additional assets in turn require more support facilities, which in turn offer more opportunities for the insurgents, and so on. Several chapters, notably 13 and 14, address ways to account for enemy behavior—to prevent enemy actions from having the effect of a positive feedback channel.

A form of positive feedback can also occur within decision making itself. For example, the HLDM issues orders to LLDMs. However, the orders happen to be erroneous and cause undesirable results. Faced with an onslaught of feedback—demands from above to explain and fix the situation, requests from LLDMs for guidance and support—the HLDM is pressed to plan and issue new sets of orders. Conceived under growing pressure, and in increasing haste, the new orders are likely to contain even more errors, in turn causing even greater deterioration of the situation, and so on [3]. Many other factors—political, psychological, environmental, local, and so on—can also introduce similar positive (that is, self-reinforcing) feedback loops.

What can be done to mitigate the possibilities and effects of positive feedback? By identifying and modeling potential sources and channels of positive feedback, some can—and should—be prevented at the design stage. Most, however, are run-time problems; that is, problems that arise only in implementation rather than design. For run-time problems, it is necessary to have mechanisms for comparing what is happening in the system with an expectation of what should be happening.

When humans perform control, and if the right information is presented in the right way, such run-time problems are often detected immediately. (See Chapter 16 for related issues.) However, when computers make control decisions—or at least some of them—it is vitally important not to lose diagnostic capabilities.

Because they are not human, computer control systems need to have monitoring and diagnostic components. Indeed, such monitoring and fault detection is becoming more common in control systems, in applications ranging from factory control to avionics. Often, it is not necessary to have a full predictive model of the system in order to detect problems. Instead, setting limits on signal energy, or noting the direction that certain signals should move in response to particular commands, may suffice. To that end, Chapter 4, for example, offers a vision of how such diagnostics can be performed with respect to enemy behavior.

SATURATION

Under certain transient conditions—and even if a system were properly sized for normal operating conditions—a C² system can overload its decision making or execution capacity of some components or links. This phenomenon—of a dynamic variable hitting a limit—is called saturation; it is one type—perhaps the simplest yet most common type—of what is known as a nonlinearity. Saturation arises when actuators, process variables, and sensors reach physical limits. Thus, when a system is linear, it is easy and intuitive to think about the effects of decisions: results (outputs) scale with actions (inputs). If, for example, one doubles the input, the output naturally doubles. In the case of saturation, however, some variable in the system has hit a limit. When it occurs, increases in the control input—which try to make the saturated variable further exceed the limit—simply have no effect. Here, the system is nonlinear, because doubling input does not double output. Although its workings are simple, saturation is subtle—because so long as variables are not at their limits, the system behaves in a perfectly linear and intuitive way.

As an illustration of saturation in C² systems, an HLDM sees an impending enemy attack in area B, and hastens to bring in assets from area A. Yet unless circumstances are favorable, he may not be able to succeed—the redeployment takes time, his logistics are already stretched to the limit, the troops are already engaged, and so on. In this case, the system's execution elements are saturated.

Saturation can also occur in a system's decision-making element: in C^2 systems, for example, human decision-makers can become saturated. It is not too difficult to imagine a case in which an HLDM has too many subordinate LLDMs (excessive span of control), which in turn will overwhelm him both with information and requests for decisions (probably in the most critical situations). Likewise, a DM may be required to have too much peer-to-peer coordination, which will similarly overwhelm him with information and requests for decisions. This sort of eventuality is also likely to happen at the most unfortunate moment, just when the DM is called upon to make critical decisions. Here, if saturation is not detected, and decisions continue to be made as if variables can go beyond their saturation limits, the overall decision-making system will malfunction. Other DMs, unaware that saturation has occurred, will continue to form their plans and expectations under the assumption of normal operations. As a result, they will not receive timely or accurate inputs from the saturated element.

One way to deal with saturation is to avoid it—never driving the system into regimes where the limits are active (by detuning the controller, for example). However, not taking a system to its limits also can mean that the system will not be operated to its full potential. Therefore, a better way to deal with saturation is to identify all limiting conditions and then design control policies for each situation. Then, having designed the system offline, it is put online, into actual operation. As saturation conditions are detected, the appropriate controller is applied. Such additional controls result in a so-called hybrid system; that is, one with both continuous—and discrete-valued state variables (see Chapter 14).

In this case, discrete state variables arise in the control logic that selects the controller depending, of course, on the state of the saturation limits (active or inactive). Since control design is much easier for each individual case, this switching control scheme offers a way, using traditional design techniques, to deal with system operating limits. When a saturation limit is active, and a control variable saturates (that is, the variable's saturation limit is reached), that variable simply becomes a constant in the system model. Chapter 11, for example, discusses an approach to analyzing and detecting limiting boundaries of different modes of system behavior.

INCONSISTENT MODELS

In hierarchical control systems, lower-level controllers are based on more detailed models of local dynamics, while higher-level controllers use models that ignore many lower-level details. Indeed, this is a principal reason for using hierarchical control architectures. If the higher-level controller used models that include all the details in the lower-level control models, a completely centralized control scheme could be implemented—and there would be no reason to allocate some decision-making responsibility to the lower-level controllers (since the higher-level controller would dictate all control). However, such a universal model—one that combines everything—is often impractical to build. Therefore, a high-level supervisory controller must be constructed with a simplified, reduced-order model.

Control theorists have developed many techniques for creating such reduced-order models of dynamic systems. Although these models include a less detailed view of the situation (that is, they focus on the so-called big picture), they nevertheless faithfully represent the system's input-output behavior from the perspective of the higher-level controller, often called the supervisor.

Similarly, a C² system's HLDM necessarily operates with models (mental, doctrinal, or computational) that involve aggregations and simplifications of details. LLDMs, to the contrary, normally use more detailed models. Indeed, there is nothing unusual or inappropriate about the HLDM using information and models at a higher-level of abstraction—and less detailed than those of LLDM. In a well-designed system, LLDMs collect information, process it, make decisions, and so on, at a level of detail appropriate to their tasks. To communicate to the HLDM, LLDMs aggregate their information into a form suitable for the HLDM. Similarly, having received orders from the HLDM, LLDMs translate their orders into more detailed actions. Clearly, there is nothing wrong with HLDMs and LLDMs operating at different levels of detail—just as long as they are able to translate correctly from one level to another.

However, such a happy state of hierarchical consistency can break down in C^2 systems. For example, in a situation when the enemy introduces new tactics or weapon systems, LLDMs continue to provide the HLDM information aggregated and summarized in the ways that once worked. Since the enemy's patterns of behavior have changed, however, the information provided to the HLDM now misses—or misrepresents—certain important variables. The HLDM, in turn, no longer has an adequate approximation of reality. Similarly, the HLDM's orders to the LLDMs no longer present all the important information—and the LLDMs' abilities to translate their orders into detailed actions are no longer assured. Here, while the world has changed in important ways, both the conventions of communications as well as the translations between the levels of hierarchy are now deficient—in subtle but important ways.

In control theory, two conceptual paradigms are used to create reduced-order models: time aggregation and state aggregation. Time aggregation refers to the elimination of fast transients not important in the supervisor's time frame. Typically, lower-level controllers operate in time-scales that are much faster than those of interest to the supervisor. The transients in these higher-speed feedback loops can then be neglected in the supervisor's model, and the supervisor can use a simpler model that focuses on the longer-term results of the lower-controllers' actions. Here, motor controllers are a concrete example of hierarchical decomposition using time aggregation. Often, in response to commands from a higher-level controller, a high-speed velocity feedback loop is implemented to control motor speed. Using a model that ignores fast transients in the motor speed, to achieve a desired task, the supervisory controller is then designed to issue speed commands to the motor. In this case, the supervisor's model simply assumes that the motor achieves the requested speed instantaneously.

State aggregation allows the supervisor to reason about system behavior, but using a much smaller set of variables. Here, the idea is to introduce a model of the system at the supervisory control level in which each state variable corresponds to a set of state variables in the actual system. For example, in the motor controller illustration, the supervisor's model may have three speeds, LOW, MEDIUM, and HIGH. Although the actual motor speed varies over a continuum of values, these three speeds may be adequate for achieving overall control system objectives.

To guarantee that reduced-order models derived for hierarchical control are sufficiently consistent with complete detailed models, various measures of fidelity are introduced, such as least-square-error or maximum error. For C² applications, reduced-order models and measures of consistency can be developed formally if detailed models of the dynamics are available (including the enemy responses). However, such reduced-order models need to be evaluated empirically, using simulation, and updated, as new information becomes available. (Obtaining good, consistent, reduced-order models for hierarchical control of complex systems remain an area of active research.) It is also possible to enforce consistency by design in complex systems, for example, by requiring lower-level decision makers to follow specific procedures that make it possible for the supervisor to predict how the system will behave without modeling the details of the lower-level controllers. For human organizational hierarchies, this is one of the most common methods for enforcing hierarchical consistency.

A majority of technologies discussed in this book rely on some type of hierarchical structures, as in Chapters 5, 8, 9, 10, and 15. All of them strive to offer various means to minimize the challenges and impacts of model inconsistencies between the layers of hierarchy.

SYNCHRONIZATION AND COORDINATION

Many distributed controllers, each responsible for regulating a particular local process variable, usually control large-scale processes, such as chemical plants. As such, a distributed architecture can have several advantages over a centralized scheme. Dedicated, single-variable controllers are easier to install, tune, and maintain. Information is obtained and used locally, avoiding the need for a high-speed communication network. The system is robust against single-point-of-failure outages—if one controller goes down, the whole process may still operate acceptably because other controllers are still performing their tasks. Such a system is not unlike a C^2 system, where controllers are multiple and distributed, and coordination and synchronization between multiple military units is necessary to achieve a common objective.

When there are multiple controllers acting on a system, however, it is possible that they can begin to work at cross-purposes, thereby leading to instabilities. An obvious example is when two units, commanded by different DMs, initiate an attack against the enemy. Unless the two attacks are coordinated in time and space, the desired impact on the enemy is lost. In all likelihood, the enemy will be able to defeat the attack, often by concentrating his assets first on one unit, then on the other.

In general, then, multiple independently executing tasks must be coordinated to achieve a synergistic effect. An example could be a series of independent attacks on a power grid. Each attack will produce a local effect; but if the time can be properly synchronized, the total effect will be a massive collapse of the entire grid. The question, then, is how can we control these actions across many organizational and command hierarchies?

In military environments, multiple-unit coordination and synchronization is achieved in part by an HLDM who plans synchronized actions, issues orders to LLDMs, and provides coordination instructions. In addition, the HLDM often issues instructions for coordination between units on a peer-to-peer basis. None of his actions are foolproof, however, because in a dynamically changing battlespace—one populated with multiple units, caught in the inevitable fog of war—LLDMs often must act only on information they have from their relatively local—and, of necessity, myopic—view of events. As but one example, LLDM A sees an opportunity to take advantage of an adversary's local weakness, thereby assuming a better position. LLDM A therefore leaves position X for position Y. At the same time, however, LLDM B, assuming it can call on LLDM A at position X if his forces need additional help, decides to take an aggressive action toward position Z. Sadly, by the time word reaches LLDM A (now at position Y) that LLDM B needs help, it is too late.

Synchronization of actions must occur not only in time but also in space. As another example, a military unit is busy erecting a tent city for refugees in location C. Meanwhile, a non-governmental medical organization builds a medical facility in location D. Heavy traffic of ill refugees now ensues between C and D, precisely across location E—which another military unit has begun to use for logistics facilities. As this example demonstrates, real world LLDMs not only do not necessarily report to the same HLDM, they also often have a variety of inconsistent—if not conflicting—agendas and objectives.

Control theory and engineering practice offer two basic methods to make certain that distributed controllers do not work at cross-purposes. The first is to introduce coordination, through either peer-to-peer communication between distributed controllers or a supervisor in a hierarchical control structure. The second is to use decentralized design techniques that guarantee the local controller actions' composite effect is never degenerative. Both approaches rely on best-attainable system models, sophisticated design techniques, and extensive off-line simulation study.

The use of coordination signals lies between centralized and decentralized design. In hierarchical schemes, the distributed controllers are designed first. Then the supervisor's control problem can be viewed as a centralized control problem, with the distributed controllers simply absorbed into the system model that is used to design the supervisory control policy. Here, the supervisor's commands are used as inputs to the distributed lower-level controllers, providing the synchronization and coordination necessary to assure that the distributed controllers work together to achieve global objectives.

Generally, implementing effective peer-to-peer communication schemes is a more difficult way to achieve coordination because it does not reduce to a classic control design problem. As a way to solve the problem, one possible scheme is to have each controller broadcast its measurements and actions to all other controllers— thereby giving all controllers global information. However, even with global information, local computations have to be designed to take into account ways that other controllers will use the information. This could lead to such extensive computations at each node—equivalent to solving the global control problem, or worse—that the advantages accruing to a distributed architecture are at least partially lost.

In current practice, communications between controllers, along with special ways for each controller to use information, are designed based on knowledge and intuition about the system. In other words, coordination schemes based on peer-to-peer communication are largely designed by application-specific ad-hoc methods, and demonstrations of their effectiveness rely on extensive analysis or off-line simulations.

Completely decentralized control strategies rely on designing the one controller's actions so that they do not adversely affect actions that will be taken by other controllers. Such design is accomplished in two ways. One approach is to design the system so that the controllers affect orthogonal aspects of the system behavior, completely eliminating interactions between controllers. The other is to design the system so that the controllers' composite actions are robust against variations that can occur in their individual behaviors, such as variations in gains or timing. Again, extensive simulation is normally required to demonstrate that distributed controllers will work well together.

Several technologies in this book offer a broad variety of approaches to coordination and synchronization, including Chapters 6 and 9—dynamic rescheduling and reallocation; Chapter 5—collaborative, radically distributed decision making; Chapter 14—game-theoretic construction of dynamic coordinated strategies; and Chapter 15—provably correct languages for coordination of multiple units.

ALLOCATING CONTROL AUTHORITY

Allocating control authority is one of the most challenging problems for designing distributed, hierarchical control systems. A large manufacturing plant, for example, has thousands of local controllers for machinery and material handling systems. While many decisions and control actions might be based entirely on local information, such as bar codes on containers, to achieve optimal performance the various processes need to be coordinated by higher-level centralized controllers. At the same time, when emergency situations occur—due to equipment failure, for example—time may not be available to communicate the situation to a centralized control system so that it can decide what to do. Indeed, it may be most expedient to respond to these situations immediately, with local decisions; communication with the central controller could occur later, and so that any effect on the overall operation could then be assessed. To design these systems, many factors need to be evaluated to determine when local controllers should make decisions and report results, as opposed to when they should report information and wait for instructions.

Designers of C^2 systems face similar challenges; Chapter 1 alludes to them, contrasting British and German approaches to command in World War I. A hierarchical C^2 system may give the LLDMs too little authority, making it impossible for them to respond to unanticipated situations in a timely manner. For example, an LLDM sees an opportunity to acquire a very advantageous position that would enable faster achievement of the goal; however, following that course would require abandoning pre-planned actions that would clearly lead to a much less attractive position. What does he do? In order to deviate from the plan, the LLDM will have to request permission. To do so of necessity would delay the planned action. In turn, delaying the planned action would result in dire consequences. Therefore, the LLDM proceeds as planned—and foregoes a better course of action (COA).

Perhaps an unforeseen situation arises that puts the LLDM in grave danger. To avoid disaster, he must take action beyond his allotted authority. While waiting for approval, time runs out. Clearly, then, allocation of authority is an age-old concern about finding the right balance between giving subordinate commanders enough autonomy while retaining the desired degree of control and safety.

Model-based design techniques have been developed to address these problems in traditional distributed dynamic systems. But the full allocation-of-authority problem remains an area of active research. Current methods focus on the operation of the system under normal conditions. For extreme conditions—those outside a normal operating regime—separate emergency control systems are designed. Because regular controllers do not have the agility or capacity to achieve what's needed in such extreme situations, it is usually not possible to use regular controllers at such times. Recent research on so-called reconfigurable control systems has only begun to consider strategies for using those distributed controllers designed for normal operation to handle new and possibly unanticipated situations. For C², such a capability would be highly desirable. For example, while local DMs certainly need the ability to act according to prescribed protocols and limits under normal conditions, they also need the agility and freedom to respond more autonomously when required to by extreme conditions. At this point, control theory may provide some analysis tools for evaluating how well the hierarchical C² system will respond to given situations. The design problems remain open.

This book presents a number of ideas that address aspects of this problem. Chapter 5 argues for radical decentralization and allocation of responsibility to an emergent collective mind of sensors and shooters. Chapter 7 offers a way to quantify the degree of flexibility and corresponding constraints that are needed in a particular COA. Chapters 9 and 10 offer approaches to negotiated allocation of responsibilities between multiple units. Finally, Chapter 15 presents a technique for rigorous design of constraints used in distributed control.

CASCADING COLLAPSE

An example of cascading collapse in traditional feedback control systems is the well-known power system blackout. Typically, blackouts occur when a single event causes a severe overload at some point in the power grid; that overload, in turn, causes equipment to fail or be taken out of service by protective relaying. Such equipment failure then causes the overload to spread to other parts of the system, thereby causing more equipment to be removed from service. The process then cascades through a large part of the power system—hence a blackout. Other networks—including telecommunications, computer, and traffic—also experience cascading failure when various arteries become blocked by congestion, leading to gridlock spread throughout the entire system. In military settings, a breach in a line of defense is one example of cascading collapse. Although only a small fraction of the overall force is defeated, the units next to the actual breach experience a great increase in the pressure applied to them. Frequently, then, the entire line begins to unravel [4].

The most effective method for avoiding cascading collapse is to keep the initial overload from spreading—for example, by absorbing its impact with local losses. In power systems, analyses of major blackouts indicate that catastrophes could have been completely circumvented: at the time of the first local overload, systems should intentionally shed a small amount of load (that is, cut off service to some customers). The difficult problem, of course, is knowing exactly when such measures need to be taken. In this book, Chapter 5 offers an extreme example of applying this idea. Other approaches to sharing the load dynamically are in Chapters 6, 9, and 10. In general, though, determining when a system is overstressed to the point of cascading collapse is both an unsolved problem and a subject of current research in the control theory community.

DEADLOCKS

Deadlock, another classic problem that has been studied extensively in the theory of concurrent DES, occurs whenever there is a circular wait for resources, a so-called deadly embrace. In a computing system, deadlock occurs when a cycle of processes shares resources—and each process holds a resource needed by the next process in the cycle. Therefore, any given process in the cycle cannot proceed because it is waiting for another process to release a resource it needs, and at the same time it is holding a resource needed by another process.

As a C^2 example, DM A requests information that must come from DM B, who in turn requests information from DM C, who requests information from DM A. However, DM A cannot respond to DM C because he is waiting for information from DM B. And so on. Here, the cycle is clearly deadlocked. A similar case would be that DM A needs resources X and Y to accomplish his mission—while DM B also needs resources X and Y to accomplish his mission. DM A acquires resource X, DM B acquires resource Y—but neither can progress farther. The system is deadlocked.

It is well known that the problem of detecting if a deadlock can occur in a given collection of processes and resources is computationally intractable; that is, the time required to solve this problem grows exponentially with the number of processes in the system. However, it is easy to prevent deadlocks by implementing an appropriate protocol for reserving and releasing resources. Therefore, in any system where it is possible that deadlock can occur, it is desirable to implement a deadlock prevention policy. However, such a policy can be expensive both to implement and execute, particularly in C². If a policy is not implemented, then deadlock detection schemes should be implemented, accompanied with procedures for eliminating deadlock conditions. Such procedures are often used in systems where the likelihood of a deadlock seems very low, and the overhead for implementing a deadlock prevention scheme is deemed not worth the effort. Dynamic renegotiation and reallocation strategies, discussed in Chapters 6, 9, and 10, are among possible practical approaches to deadlock prevention.

THRASHING

Thrashing describes a number of different phenomena that manifest themselves in a similar manner: a system undergoes multiple unproductive cycles of behavior.

Chattering, one type of thrashing found in traditional switching control systems, arises when an action taken after a switching condition is reached (typically when some signal crosses a threshold) drives the system quickly back to the condition before the switch occurred. Sometimes, cycling between two control modes is desired—for example, when a furnace thermostat detects that the temperature has gone below the specified set point, the action (turning on the furnace) returns the temperature above the threshold (where the furnace shuts off). Although a thermostat is designed to cycle, it would be inappropriate if it switched the furnace on and off every 10 seconds, as might happen if the thermostat is too sensitive to temperature changes. In this case, it might be said that the thermostat controller is chattering.

Chattering in a C^2 context might occur, for example, when reserve forces are being sent to provide help to two different battle sites. A naive policy for deploying the forces would be to send them immediately to the site that is presently in the most need. However, when the forces arrive at one site, the other immediately becomes more in need of the extra forces. Such a policy would lead to repeated reallocations of the reserve forces, consuming time and assets without garnering productive results.

Generally, a C^2 system may perform unnecessary switching—from task to task, or from plan to plan—to the point that excessive resources are used up simply by switching, and not by any productive effort. Even though the causes of such continuous changes of plans or tasks may be perfectly reasonable—the arrival of new information, for example—such thrashing can be extremely disruptive and disorienting, particularly for lower-level units [5].

In control engineering, the possibility of thrashing can be detected by analyzing system behavior at critical switching points. (Global simulation is not necessary.) When thrashing conditions are detected, two things can be done to reduce a potential harmful effect. First, it may be that simple modifications in the switching logic can be implemented to keep switching frequencies within acceptable limits. One such modification creates hysteresis in the switching by introducing a gap between switching thresholds in opposite directions. This is precisely what is done in a thermostat to keep a furnace from cycling on and off too frequently. Another modification is to implement dwell time, requiring the system to remain in any control mode for a minimum amount of time.

The second thing that can be done to reduce thrashing's potential harmful effects is to introduce a cost for switching, an approach applicable to controllers making decisions based on optimization. If switching does have a cost, a controller will not switch until other factors have changed sufficiently to compensate for the switching cost. Indeed, in many applications undesirable thrashing occurs because the cost of switching has been overlooked in controller design. For example, in a traditional control system excessive switching might cause harmful wear to the equipment. In C² systems, thrashing will probably be minimized if the true cost of a switching decision—moving troops, for example—is assessed correctly. By incorporating such a cost, one achieves a dwell time that reflects the system's true operating costs and objectives. The technologies discussed in Chapters 6 and 7, for example, introduce special mechanisms for imposing penalties for changes in plans, thereby minimizing unnecessary thrashing.

Livelock, another thrashing behavior, occurs when processes appear to be making progress at each decision point, but are actually stuck in an unproductive cycle of steps. For example, in state A, the DM sees path A-B-C as the most appropriate toward the goal, and so he moves to B. From that vantage, path B-C-D looks optimal, so he moves to C. From there, path C-D-A looks optimal—and so on. Here, either cyclical or non-cyclical non-convergent behavior is possible.

In a system of multiple decision-makers, a process of negotiations between decision-makers can also cause a pattern of unproductive, thrashing cycles. In a sequence of events between two DMs, (i)

DM A recommends decision X1 to DM B; (ii) B responds with recommendation X2 to A; (iii) A recommends X3 to B; (iv) B recommends X4 to A; then (i) again, and the chain repeats—livelock [6]. In particular, negotiation-based technologies discussed in Chapters 9 and 10 include mechanisms for preventing both deadlock and livelock.

In general, to check if there is some possibility for livelock in a set of concurrent processes, formal analysis methods can be applied to graphical representations of the process cycles. The formalism of Petri nets is particularly useful for this purpose. When there are several components in the system, simulation is a less attractive method for trying to detect livelock conditions—because it is usually impractical to simulate all possible execution sequences.

Livelock can be avoided by not permitting any processes in the system to execute cycles indefinitely. Local monitoring is sufficient to determine when a cycle has been entered, or when no progress is being made after a given number of steps. When no progress occurs, a procedure must be available for getting the system out of the situation causing livelock. For example, a protocol might be proscribed that selects one process to take unilateral action, allowing the process to progress independently and without regard for what other processes might be doing. In general, such procedures are application-specific, designed based on insights into a system's properties and overall control objectives. Here, formal verification can be used to assure that the procedures indeed eliminate the possibility of livelock. Technologies discussed in Chapters 14 and 15 include certain types of such formal verification.

SUMMARY

Failures or undesirable pathological behaviors in complex systems are studied in control theory and computer science; such studies may also offer insights and solutions for C^2 systems. Many complex system behaviors are not unlike those in C^2.

Time delays between an event's occurrence and a corrective action can cause instabilities—counterintuitive and violent divergence from normal behavior. Control engineers use de-tuning or predictive techniques to mitigate time delays.

A controller with excessive gain is analogous to an overly aggressive commander, while a controller with low gain is like an overly cautious commander. Control theory offers methods for determining the right gain for a given situation.

Positive feedback is a class of instability in which corrective action makes a situation worse rather than better. Saturation, an overload of a system component or link, causes counterintuitive and unexpected effects. It is possible, however, to design an effective controller that accounts for saturation.

In a hierarchical control system, inconsistency between models at different levels can lead to wrong decisions. Rigorous methods—building reduced-order models—exist for reconciling such inconsistencies. Loss of coordination and synchronization is common in distributed control architectures as well as in C². Formal techniques exist to design coordination schemes.

Allocation of control authority—too little or too much local autonomy—leads to tradeoffs between safety and agility. Control theory offers techniques to analyze such allocations. Deadlock is a deadly embrace in which several entities hold up each other's resources. Thrashing and livelock—repetitive patterns of unproductive actions—are common in control and computing systems, as well as in management and C². Methods are available to predict and to avoid thrashing even in very complex systems.

PART II
MILITARY INTELLIGENCE

CHAPTER 4

DISCERNING THE CODE
OF ENEMY INTENT

On October 8, 1806, Napoléon receives news of the first engagement of what would later be known as the Jena Campaign: Murat's vanguard defeated a Prussian force under Tauenzien [1]. Combining the eyewitness report of his adjutant with those of prisoners, the Emperor forms his estimate of enemy disposition and intent: he believes that Prussians will attack his left flank. Just a few hours later, upon receiving another dispatch, he revises his estimate dramatically: no, the Prussians are retreating toward Gera and will assemble there. Based on this, by October 11 the Grande Armée concentrates its forces around Gera, only to discover that the Prussians are nowhere to be found. Uncertainty reigns for more than a day, until reports from Soult and Murat reveal that the enemy is far to the west of Gera and is retreating, possibly to the north.

Based on the new information, the Emperor issues a series of orders, which executed with exceptional rapidity, reposition the French forces and cut off the Prussians' retreat. By late October 12, a series of additional reports bring the Emperor his long-desired clarity. Yes, the enemy is retreating north, toward Magdeburg, or so Napoléon assures his lieutenants. Yet, by the late afternoon of October 13, riding towards Jena and observing an attack of Prussian forces on the troops of Lannes, Napoléon once again reverses his estimates and concludes that the entire Prussian army is engaged at Jena. Yet his eyes deceive him, and he is mistaken. Facing him at

Jena is merely a flank guard under Hohenlohe. Unaware of the truth, Napoléon personally directs divisions and even regiments, completely convinced that he is fighting the main action of the campaign. The battle of Jena rages throughout October 14.

Not until late that day, when the battle is won, does the Emperor discover that truth which he initially refuses to believe. His fight at Jena was merely a sideshow, for the main Prussian army is at a different location and retreating to the northeast. Fortunately for the Emperor, and unbeknown to him, the Grande Armée 3rd Corps stumbled across the retreating enemy at Auerstadt and managed to defeat it.

Although conducting a successful campaign against a less capable opponent, Napoléon was unable to achieve a clear and accurate understanding of the enemy's intent. Even an outstanding, victorious military leader, who has brilliantly fought many campaigns, finds the challenge of identifying enemy intent immensely difficult.

THE QUEST FOR A CRYSTAL BALL

A dream of every decision-maker operating in a competitive environment is to know the intent and the likely actions of his competitor. A president of a company will give anything to know how the competition will respond to the introduction of a new product. A commander of a military operation will go to great lengths to uncover the location of an impending enemy attack.

In today's military C^2, the task of identifying enemy intent is handled by highly specialized organizations, which collect information in support of a commander's requirements. Once collected, intelligence organizations process the information and advise the commander on what appears to be the most likely enemy intent and disposition, including likely courses of action and the most dangerous responses to various situations.

These organizations certainly do not have a crystal ball, and their proverbial bag of tricks does not include a magic device that might predict enemy actions based on what the enemy has done in the past. Yet scientists and technologists are working on tools that seem to offer capabilities that approach the legendary crystal ball. One such technology is demonstrated by OpSpy [2], which has been developed by a team of researchers from the Massachusetts Institute of Technology, Laboratory for Information & Decision Systems, Cambridge, Massachusetts, and the University of California Los Angeles, Department of Mechanical & Aerospace Engineering, Los Angeles, California.

THE DNA OF ENEMY BEHAVIOR

The complex factors that comprise and affect a military force's actions are enormous, ranging from political to logistical, weather characteristics to local population, staff training to medical services, the highest commander's personality to squabbles between officers to each private's psychology. These factors are contradictory, fluid, and often intangible. Certainly, it is a technologist's dream to capture this universe of issues in a neat, manageable package of numbers, but is it possible?

One unlikely yet suitable analogy is DNA—the complex code that governs the development of biological organisms. An enormously complex structure such as a human body, and the sophisticated processes of its development and growth and functioning, are encoded in a relatively small (although still very intricate) package that in recent years has been studied and understood with greater and greater ease. The OpSpy idea is somewhat similar—to distill the sum total of an infinite variety of factors affecting a military force's decisions and actions into a relatively small collection of underlying numbers. This kind of DNA, invisible and perhaps unrecognized by anyone within the military organization, governs its overall behavior. Is such a code really there? The OpSpy researchers argue that such a code does exist and can be computed from a military force's observed behavior.

By analyzing the enemy's past actions, OpSpy computes a mathematical image of enemy intent—a reflection of enemy priorities and objectives, and the enemy's preferences in achieving them. Such a code of intent in itself is neither particularly comprehensible nor useful to a human analyst. However, by combining this code with an intended or hypothetical friendly course of action (FCOA), the computational tool can predict how the enemy will react to a given sequence of friendly actions. Such a prediction can be presented to the human analyst in a number of ways, including, for example, as a graphical display of enemy movements and actions. OpSpy, therefore, does act like the mythical crystal ball, predicting future enemy actions based on the enemy's own recent history.

This technology's likely user would be an intelligence analyst. Taking army operations as a specific example, we may envision that such a capability could be available to Division analysts. Receiving a request from a Divisional Planning Staff for an updated analysis of enemy course of action (ECOA), an analyst would use OpSpy on a computer, or portable terminal connected to a reach-back computing center, connecting with a Situation database to extract records of the enemy's key past actions. OpSpy further uses the harvested information to compute the enemy's code of intent, then applies it against the analyst-specified FCOA. The result is several possible ECOAs, each with an associated measure of likelihood. The analyst then views the graphic representations of alternative ECOAs—animated troops movements, for example—and makes conclusions about the most likely and most dangerous ECOAs.

FIVE HOURS IN THE LIFE OF OPSPY

Exploring OpSpy's use in more detail, and in a more realistic setting, let us consider a hypothetical scenario pictured in Figure 4-1.

H hour to H+48:00

Red combined forces invade their smaller and weaker southern neighbor, Green. Red has long coveted Green's high tech resources, large-capacity oil and gas refineries, and extensive oil and gas reserves. The Red attack is in Corps strength, with mechanized infantry and armor units supported by airborne assault and close air support across a 200-km front. Simultaneously, Red air forces attack strategic targets within Green, and Red naval forces blockade Green's only port. The attack is unexpected, and preparations have been well concealed. Green defends valiantly with inferior ground and air forces, but is unable to stop Red's inexorable advance.

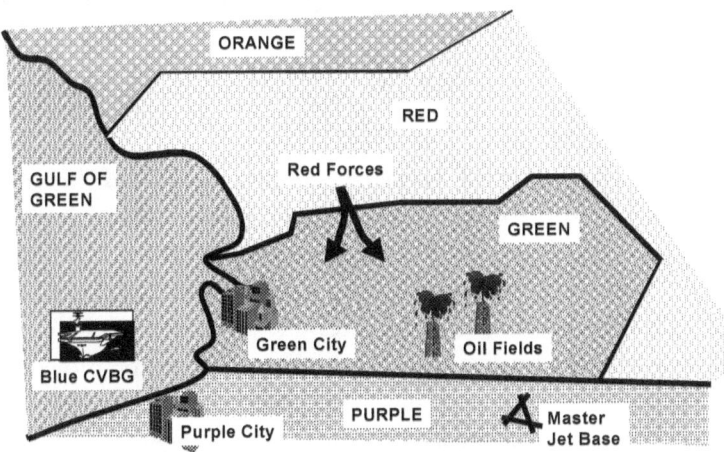

Figure 4-1. This map depicts the battlefield situation discussed in the OpSpy example.

Green immediately requests help from its distant ally, Blue, which has limited, but potent, naval forces in the theater, including a carrier battle group and a Marine amphibious ready group (MARG). Blue National Command Authority responds quickly by executing OPLAN 8332, which steams naval combat forces towards the conflict area and prepares to launch fighter sweeps and interdiction missions against Red forces. The plan also initiates the immediate deployment of theater air forces and CONUS-based ground forces into Purple, a large friendly country adjacent to Green's southern border. OPLAN 8332 gives total operational authority to the U.S. Joint Commander in Chief (CINC) in the theater to employ, as he sees fit, any and all conventional forces within his purview.

Within 36 hours, the combined effects of the Green tactical withdrawal, along with Blue air and naval interdiction, have slowed Red's advance, but it appears that Green's capital will be overrun before sufficient forces are available to stop the invasion. Adding to the Red potential is a large indigenous population of ethnic Reds within Green territory who are considered to be organized and sympathetic to the Red cause.

H+48:00

Completely unanticipated by the Allied command, Red initiates another attack to the north against Orange, also home to a large Red ethnic population. Although the attack is not of the same magnitude, in terms of quantity of forces deployed, this second front is well inland, not nearly as accessible to Blue naval forces or other reinforcements. Further, this development is not provided for in the OPLAN, so that any force movement will be ad hoc until the plan can be revised. The province of Orange, at the point of invasion, is largely barren terrain with no proven oil reserves or other resources. While the Red attack will surely bring Orange forces into the Allied coalition, those forces are not forward deployed, nor are they integrated into the Allied command hierarchy.

The CINC now realizes that it is not at all obvious what Red is actually trying to achieve. Indeed, Red's objectives, intents, and priorities can be interpreted in several significantly different ways. As such, the CINC is concerned that his most recent request, to airlift his reserve ground forces into Orange to counter this latest attack, may not be his best option. In fact, all the types and locations of his requested actions may not be optimum, depending on Red's true intent.

It is in just this type of situation that OpSpy can help identify Red's intent and probable COA.

The Blue staff OpSpy systems operator, an intelligence officer working with support from the reach-back center in CONUS, notices that one of Red's key objectives seems to be encouraging its own ethnic minorities within the now-occupied territories to support the two invasions. Since this type of objective is not in OpSpy's latest database, it therefore needs to be manually entered. Taking this data into account, the operator completes the entry of new types of Red forces and objectives into his OpSpy tool.

H+48:10

The operator begins to form a hypothesis, a collection of objectives that the enemy is likely to pursue. By providing OpSpy with one or more hypotheses, the operator helps the tool formulate the overall cost function (to be discussed later) that the enemy optimizes in conducting his operations.

Based on what has been observed in the last 48 hours, and after entering his new data, the operator starts to produce a predicted ECOA update for the next 120 hours. To help OpSpy focus, he specifies several of the most likely tangible and intangible objectives that the enemy will likely pursue, including:

— Destruction of or disabling Green and Orange air defense systems, including Green air forces and their bases
— Destruction or capture of Green ground forces

— Capture or denial of Green and Orange communications capabilities, both political and military, within its occupied territory
— Occupation of Green's major oilfields and refineries near SomeTown
— Imposition of psychological warfare on Green's population, in part by encouraging ethnic Red minority populations to rise up and join the Red invasion against Green and Orange

H+48:20

The operator also uses OpSpy's graphical user interface (GUI) both to enter FCOAs and specify constraints to those COAs. Here, for the purposes of OpSpy, each COA is described as a temporal sequence of states representing battlefield actions or events from the current time to a future point. In general, COAs are available as online files or database records (which represent the COA in a formal language understood by a computer).

The operator imports these COAs into the OpSpy, including all FCOAs corresponding to OPLAN 8332, and revised FCOAs based on latest enemy movements on the new, Orange front. In particular, the FCOAs require deployment of the Fifth Mountain Infantry Division, earmarked as Blue ground force reserve in OPLAN 8332, to bases in Orange to assist Orange forces counter Red's incursion into Orange territory.

The operator also uses the OpSpy GUI to input constraints¾observed and deduced enemy parameters culled from a variety of intelligence sources. OpSpy will then use such constraints to eliminate unreasonable solutions from its computations, for example, a solution where an enemy unit moves from one location to another at an unrealistically high speed. The constraints include such estimates as:

— Enemy strength within the battlespace for of force

— Enemy speed of movements
— Time required for enemy transition from one disposition into another

H+48:40

OpSpy extracts information from the SDB that records enemy past actions in the area of interest. Here, OpSpy produces a temporal sequence of enemy observations, from a certain point in the past to the current moment—in this case, starting with the initial H-hour invasion of Green. In addition, the friendly forces' past actions are also defined. As OpSpy defines it, a state includes, in each geographic area, enemy strength for each force type, as well as force dispositions. A typical readout might be, for example, "at time T25, in quadrant ABC, there is one battalion of mechanized infantry in prepared defensive positions and one battery of SAM-X on the move."

H+48:45

After the operator specifies several FCOAs, OpSpy computes enemy intent and then applies it against the FCOAs—an intricate computational process. Comparing a tree of all possible actions that the enemy could have taken with the path that he has actually taken, OpSpy computes the relative values that the enemy implicitly applies to each action. Then enemy intent—the collection of all these values—is used to determine which future actions the enemy is likely to prefer.

As a result, OpSpy computes several possible ECOAs, each with an associated measure of likelihood. In this case, OpSpy comes up with and presents via GUI three distinct possible ECOAs, while a movie-like display makes it easy to interpret and follow the battle's evolution over the next 120 hours.

Unfortunately, due to the SDB's relatively incomplete nature—

because it contains only data from the previous 48 hours—OpSpy is not very certain about which of these ECOAs is most likely:

— The first, with a .36 probability, states that Red will continue its thrust southward to Green's southern border, then dig in.
— The second, at a .35 probability, is that Red will continue to SomeTown to overtake the oil and gas fields.
— The third, at a .29 probability, is that Red will pull back and consolidate to a perimeter that includes both the preponderance of Green refinement capacity and the ethnic Red population.

H+49:00

The operator directs OpSpy to suggest friendly actions that would rapidly and efficiently provide additional information to help determine which of the three ECOAs is most likely. In response, OpSpy proposes several defensive counterstrikes in key areas that would likely force direct enemy responses. Then, observation and analysis of these responses may provide OpSpy with enough data to deduce more concrete enemy intent.

To arrive to such recommendations, OpSpy performs another complicated computation, in essence exploring the tree of the enemy's possible actions and trying to impose additional actions that would help to differentiate between objective values within the hypothesis. The computation results in OpSpy proposing probing actions designed to elicit telling enemy responses. In this case, such actions would include:

— Reinforced battalion-strength reconnaissance-in-force— frontline Purple mechanized infantry supported by on-call close air support—into the enemy's right (western) flank
— Rapid maneuver of the MARG and supporting combatants

into position for an amphibious assault adjacent to Green's port—the action designed, at this point, to be a feint

— Engagement of Red surface units blockading Green's port by elements of the carrier battle group

— Preplanned naval cruise missile attack against military and industrial targets in Red's capital city

H+49:15

Through the staff director of intelligence (G-2), the operator submits the proposed probing actions to the CINC. Although OpSpy's inner models and workings are complex and abstract, computational process conclusions are presented in an operationally relevant, clearly recognizable, easily visualized form. Of course, there is an OpSpy display in the war room enabling the CINC and his staff to digest the proposals in the same user-friendly format as the operator.

H+49:45

After carefully considering the proposal, CINC, in consultation with his top advisors, elects to proceed with the recommended actions, in addition to continuing support of the Green force withdrawal. He agrees with OpSpy that any enemy responses will offer insight into Red's true intent.

H+50:45

Ships in the Blue carrier battle group launch a salvo of 24 Tomahawks towards the Red capital. With its escorts, the MARG continues at flank speed towards the Green port, helo rotors churning on deck, all of it easily seen by Red sensors. The Purple infantry thrust into the Red flank lands its first rounds on Red forces. On the floor of the United Nations in New York, seven ambassadors line up to denounce the Red invasions.

H+51:45

Having proposed specific probing actions and the likely range of enemy responses, OpSpy helps focus the collection's requirements, reducing overall load on military intelligence assets and improving their resulting output quality. Here, reports of enemy responses to the probes have been arriving from a variety of automated data sources and electronic reporting systems, and accumulating within the OpSpy SDB. Across the southern front, Red forces have stopped their advance and appear to be digging in to defensive positions. Red combat air patrols have increased. Red air and artillery have engaged units of the MARG. Red forces in the north appear to be withdrawing from advanced positions in Orange. The Red propaganda machine denounces the Allied cruise missile attack as a war crime against innocent non-combatants, claiming 42 civilian casualties.

H+51:55

A typical analysis cycle is roughly 10-20 minutes, primarily dictated by the operator's need to select enemy objectives and then review the results of OpSpy's computations. At this point, using the updated SDB, the operator reruns OpSpy—which produces far more definitive predictions. Now the most likely ECOA, that Red will dig in as it approaches Green's southern border, is a .63 probability, far ahead of its closest competitors. Likewise, the probability that the enemy will continue its northern incursion is now near zero, indicating that the Orange invasion was indeed diversionary.

H+52:00

Using a rigorous model, a range of ECOAs consistent with enemy intent derived from observed enemy actions has been computed and made available for commander decision making. Now, the OpSpy operator and G-2 team study predictions of specific enemy actions 120 hours in the future.

It now becomes clear that the OPLAN 8332 ground force reserve (Fifth Mountain Division) deployment to the northern front is not the best option. Indeed, not only would those troops be of little strategic value, but their absence in the South would also deprive the Allies of their planned reserve. Based on Red's response to Blue actions, OpSpy's predictions call for a different FCOA that justifies redirecting the Fifth Mountain to bases in Purple.

H+52:30

Derived systematically from enemy intent, weaknesses in the currently accepted FCOA are now objectively identified. Alternative FCOAs are proposed, and G-2 and the OpSpy operator present the new results to CINC and his advisors in the war room.

H+53:00

CINC approves new orders, changing the priorities of the airlift. The operator returns to his duties. OpSpy is ready to process the latest reports from the front and produce a new update of enemy intent.

THE SITUATION DATABASE: HOW MUCH IS ENOUGH?

To see the inner workings of OpSpy, let's begin with OpSpy's data input. Of paramount importance, the SDB must contain the following information, or the equivalent thereof:

— Enemy past actions—a temporal sequence of enemy state observations, from a certain point in the past to the current moment. As mentioned earlier, a state is a collection of information such as strength of enemy forces of each type and the disposition of the forces, specified for each geographic area of interest.

— Past actions of friendly assets—a temporal sequence of friendly states.

— Key characteristics of various battlespace zones or areas, including terrain, airspace, and weather. Most of these are time-dependent and therefore must be described for multiple time periods.

Here, SDB materials are not raw situation reports or imagery, but instead have been pre-processed—exploited—by appropriate systems and analysts.

Overall, the hypothetical SDB's contents are not unlike information typically collected by today's military intelligence systems and organizations, and the SDB itself is not unlike today's intelligence information databases. An example is the All Source Analysis System (ASAS), the U.S. Army's battle command intelligence system that provides ground enemy situation.

With regard to such databases, one potential benefit of OpSpy's rigorous mathematical modeling is that it helps military intelligence systems focus on key information. Rather than broadly tasking military intelligence for situation awareness, for example, the OpSpy model instead offers guidance on specific tasks keyed to critical information. And rather than generically asking for everything, our analyst and his tool ask for just what is needed—and no more. Indeed, if nothing else, the OpSpy approach has the potential to reduce significantly the total load on military intelligence resources.

Of course, one might be concerned about SDB size, but it need not be excessive. Here, for example, is one possible rough estimate of the information required for each analysis. Let's assume that there are 100 significant asset units, both friendly and enemy, and 20 data items—location, strength, posture, speed, direction, and so on. There are 50 time points needed to perform the analysis, approximately every hour for the last 48 hours. Therefore, the data items total will be 100*20*50 = 100,000. Indeed, numbers of such magnitude are well within the capabilities of modern databases.

To execute its analysis, OpSpy initiates and performs SDB information retrieval in "pull" fashion, on average every half-hour, while SDB updates occur in "push" fashion, with new, multiple-source electronic reports arriving as often as every few seconds. (Such numbers are, of course, merely illustrative and may differ depending on circumstances.)

Of course, SDB information accuracy is not going to be perfect, for all real-world data suffers from latency, uncertainty, and incompleteness. In addition, it contains a certain percentage that is outright erroneous, due to many factors, not the least of which is enemy deception. Such data imperfections will certainly reduce the quality of OpSpy's analytical results—but the imperfections will not defeat the results entirely. Instead, quality degradation will be gradual—while OpSpy, to some extent, is actually able to correct or filter out input data inaccuracies.

THE HUMAN TOUCH

While the SDB brings the most voluminous data to OpSpy, OpSpy's human operator provides other information. Using a GUI, the analyst inputs hypothetical future FCOAs, with each FCOA a sequence of states from the current time point to a future point. An analyst may enter the FCOA in one of several ways, including receiving an online file in formal language and importing it into OpSpy, sketching the FCOA using an intelligent graphical interface which automatically translates into formal computer language; or using a combination of the two.

The analyst also enters a collection of constraints, particularly constraints on:

— Total enemy strengths within the battlespace for each type of force (possibly a time—variant constraint)
— Movement speed (depending on force type and terrain characteristics)

— The time required to transition from one disposition into another (pair-wise for dispositions)

Such constraints are important because they allow OpSpy to construct additional equations that guide its algorithm away from unrealistic, physically impossible solutions.

Another input the analyst provides is a hypothesis about the enemy's likely objectives and preferences, including capturing certain friendly territory, destroying certain friendly assets, preserving certain enemy assets, and so on. Entering such a hypothesis can be a point-and-click exercise, from either a list of stored objectives and preferences or a direct entry of new or unique objectives. In any case, such a combination of objectives and preferences allows OpSpy to construct the cost function that the enemy attempts to minimize.

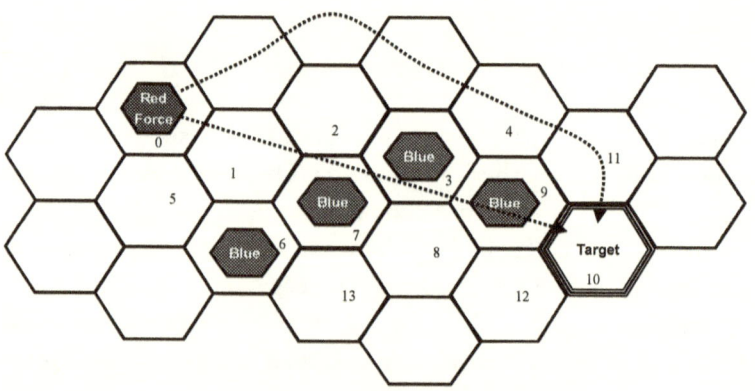

Figure 4-2. A Red force attempts to approach the target while avoiding the Blue forces. The Red commander faces tradeoffs between the speed of the operation and its risks.

Consider the example above where an enemy Special Operations Forces unit is advancing toward an objective through terrain that offers excellent cover and concealment. Blue believes that the enemy values the speed with which the operation is executed, and also wants to avoid detection by and engagement with Blue forces, which could reasonably destroy Red's elite SOF assets.

Such a scenario can be expressed mathematically: the enemy is in effect trying to minimize the cost function $F = C1*\text{Time-of-movement} + C2*\text{Risk-of-detection} + C3*\text{Attrition-of-force}$. Here, the coefficients C1, C2, and C3 can be considered relative values or costs that the enemy implicitly assigns to his cost-function components, and these coefficients' set of values is precisely the previously discussed DNA of enemy intent, reflecting the enemy's total preferences and goals. OpSpy's ultimate job, therefore, is finding the values of coefficients C1, C2, and so on. However, based on his experience and judgment, the human analyst is best able to guess and specify the form of the cost function—that it includes a combination of speed and detection risk, for example.

Do we seriously believe that enemy actions are guided by a rational optimization of some well-defined and mathematically formulated cost function? Have we never heard of the infamous fog of war? Of plans that become irrelevant even before an operation starts? Of irrational or seemingly irrational acts that so often decide a battle's outcome? Needless to say, OpSpy researchers are not unmindful of such issues, and they understand that a real-world commander rarely formulates his goals, objectives, and relative values with mathematical precision. The researchers believe, however, that all such complex, often irrational, and seemingly un-mathematical factors do combine in a way that can be approximated by optimizing a certain cost function.

To help the analyst pick the appropriate hypothesis components, OpSpy offers a list of possible objectives, some generic ("minimize attrition of friendly forces," for example), others more specific and formulated about a specific situation ("destroy port Dikturi"). The analyst then faces an interesting tradeoff in picking the hypothesis' components. By choosing fewer components, he risks misleading

OpSpy into focusing on the wrong aspects of enemy intent. But by playing it safe and picking a large number of possible components, he leads OpSpy to produce a less definitive, more vague answer.

The decision on when to invoke OpSpy is also an analyst's input. For example, the analyst could invoke OpSpy at regularly scheduled intervals, or upon the arrival of significant new information about enemy actions. Naturally, an analyst will also invoke the OpSpy for a new staff request. And it is also possible to use an autonomous operation mode in which OpSpy automatically monitors the SDB and alerts the analyst when the enemy actions begin to deviate significantly from the most recently computed intent.

WHITHER THE TWAIN: ENEMY COA AND FRIENDLY COA

Ultimately, OpSpy's first and most important job is to compute enemy intent, represented as coefficient values associated with each hypothesized cost function component. Once again, a particular intent could be described using coefficient C1 (the importance attached to minimizing movement time to target) of .85, coefficient C2 (the importance of avoiding detection) of .63, and coefficient C3 (force preservation) of .07.

With intent stored internally for immediate reuse (to analyze another ECOA, for example, without recomputing intent, which is relatively expensive), OpSpy also archives it for after-action review.

In this form, however, the intent is not particularly comprehensible even to a well-trained and experienced analyst. OpSpy's collection of dry numbers needs further processing—it must be associated with clear predictions of the enemy's tangible future actions.

Another OpSpy output is one or several ECOAs for each given FCOA—a prediction of future enemy actions based on identified enemy intent. OpSpy then stores these FCOA-ECOA pairs for immediate reuse (including analyst review), and also selectively archives them for future review.

Since a picture is worth a thousand electronic records, as an analyst wants to review and picks from computed ECOAs, OpSpy presents a visualization of ECOA-FCOA pairs. One example might be a movie-like animation—movements of unit icons on a terrain map. Obviously, more sophisticated visualizations may be possible as well.

FROM THE FOREST OF OBSERVATIONS TO THE TREE OF INTENT

OpSpy begins its analysis by computing enemy intent—with the analyst hypothesizing the cost-function coefficient values. Greatly simplified, the process's key steps are [3]:

1) OpSpy extracts data from the SDB and forms a tree (or, more accurately, a network) of possible enemy actions. For example, Figure 4-2 depicts a simple case in which a Red force intends to approach its target covertly by moving over terrain that offers good cover and concealment, while several Blue units intend to intercept the Red attacker.

2) OpSpy uses SDB data to formulate a network of possible Red force actions, here depicted in Figure 4-3. The analyst enters his input, specifying the cost-function form: F = C1*Time-of-movement + C2*Risk-of-detection + C3*Attrition-of-force. Depending on the relative importance that the Red commander assigns to these different considerations—coefficient values C1, C2, and C3—the Red unit may pursue different routes. The analyst uses previously entered constraints to exclude impossible actions and herefore limit network size. For example, the analyst may have determined that it would be extremely unlikely for the Red force to traverse zone 13, in Figure 4-2, and therefore

he excludes that zone and any associated movements in it from the network in Figure 4-3. Each movement is associated with expenditures of certain time (different due to different terrain) and with a certain risk of being detected and engaged by the Blue security forces. Data is shown for only three of the movements to reduce the clutter.

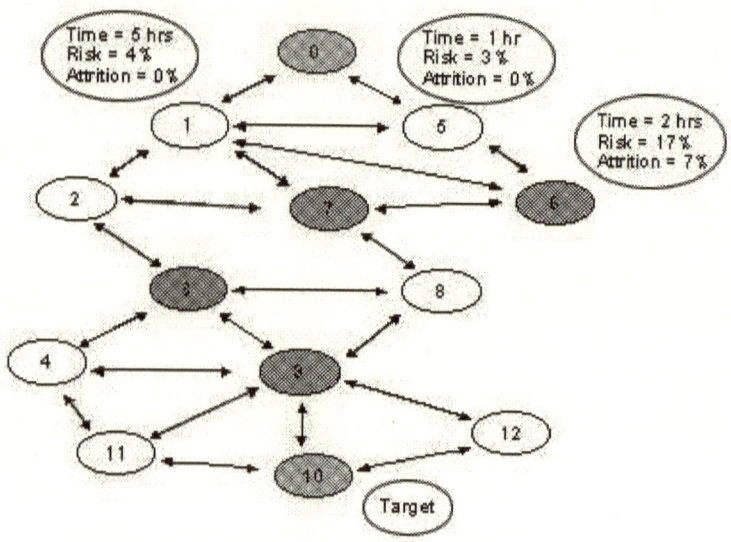

Figure 4-3. This network depicts the possible actions of the Red unit—movements from zone to zone.

3) OpSpy formulates a number of equations corresponding to each network node. Consider Figure 4-3. Let's use Lowest-Cost-0 to represent the lowest cost at which the Red unit can move from Zone-0 to Zone-10 (the target). Similarly, Lowest-Cost-5 denotes the lowest cost to get from Zone-5 to Zone-10. Now, Lowest-Cost-0 has to be Lowest-Cost-5 plus the cost of getting from Zone-0 to Zone-5, which means to expend one hour, accept 3% risk, and suffer 0% attrition. In other words, Lower-Cost-0 = Lowest-Cost-5 + C1*1hr + C2*3% + C3*0%. Similar equations can be formulated for

other nodes in our network—and OpSpy can use them to find unknown coefficients C1, C2, and C3.

4) OpSpy also formulates inequalities corresponding to each node where the enemy has made a known choice among possible alternative actions. Suppose, for example, that the Red unit has already made moves from Zone-0 to Zone-1. Why did it do that? Why did it not go from Zone-0 to Zone-5? After all, it takes much longer to get to Zone-1 than to Zone-5—and the risk is higher. Well, the enemy obviously feels that there will be a lower cost getting to the target from Zone-1 than from Zone-5, and that lower cost will offset the additional four hours spent getting to Zone-1. Mathematically, OpSpy can write an inequality connecting risks, times, and so on, to the coefficients and lowest costs. Similarly, it can write an inequality for each node where the enemy has already made a choice.

5) OpSpy takes the previously derived equations and inequalities and solves this large system. Ideally, the resulting solution will render all unknowns¾coefficients C1 and C2, Lowest-Cost-1, and so on. In reality, the information available in this system of equations and inequalities is not usually sufficient to find exact values of coefficients. Instead, the system only finds a numerical interval to which each coefficient belongs. For example, OpSpy might say that C1 is larger than .6 but less than .8. Of course, the more components the analyst includes in his hypothesis of enemy objectives, the more coefficients—and the less opportunity to narrow the values' interval for each coefficient. Similarly, the fewer enemy action observations, the less information available to define precise intent. In any event, there is at least some information about enemy cost function coefficient values. In other words, OpSpy is presenting at least a partial understanding of enemy intent.

Predicting Enemy COAs

Once intent is determined, and cost function is known (at least to the degree that we know coefficient intervals), OpSpy can predict future enemy actions. Building a tree of possible enemy actions, at each node (the choice between alternative actions) OpSpy uses the cost function to determine which enemy action is more likely. We do not need to explore this process in more detail here. There are a variety of ways to accomplish such predictive planning, and some of them are discussed, for example, in Chapters 12 and 13.

We also do not have the space to discuss another aspect of OpSpy computations—the ability to determine suitable friendly probing actions (recall the recommendations OpSpy made at H+49:00).

A Simple Deduction

The earlier statement about the tree of all possible actions must raise concerns about OpSpy's computational feasibility, for it is easy to see an astronomical number of possible permutations, of innumerable situations and actions; and, equally, of corresponding astronomical computation times required to sort through these infinite possibilities. Clearly, such algorithms can indeed have extremely high computational complexity, vastly exceeding any computer capabilities even imaginatively existing many years from now.

However, OpSpy researchers found an efficient way to reduce the computational problem to a manageable size. Their algorithm takes an amount of time that is linear in terms of observation space—the size of each battlespace snapshot times the number of such snapshots in the SDB. If the SDB (or at least a subset useful for analysis) has 100,000 data items, then the number of equations to be solved will be at most of similar magnitude and not too huge a number for modern computers and algorithms. Besides, eliminating impossible situations via application of constraints we discussed earlier will drastically reduce this number.

YES, BUT . . .

OpSpy still faces a number of very serious challenges. If unpredictability is one of the fundamental tenets of warfare, it may seem counter intuitive that a machine can predict what a wily enemy commander might do. In modern simulator-based war game scenarios, players are routinely able to outmaneuver the enemy simply by doing the opposite of what they are expected to do. As a prediction tool, for OpSpy to succeed, it must cope not only with randomness, but also with the enemy's distinctly unpredictable nature. Will OpSpy's predictions make our analyst, and the commander he supports, too predictable to the enemy? One answer is that ultimately it remains for the intelligent commander to make himself unpredictable to the enemy no matter what any analytical tools might suggest.

The inherent discontinuity in military objectives and intent provides another conceptual challenge to OpSpy. Indeed, such significant changes are a reality of many—if not all—military operations. Typical examples include exploiting an unexpected opportunity, collapsing morale when self-preservation becomes the overriding factor in troop behavior, and so on. It is also common to plan for priority changes from one operational phase to another. Following our earlier scenario, the enemy commander may plan to preserve his armored corps during the first phase, but may then employ that force decisively—and with little concern for attrition—in the next phase. Such discontinuities may render OpSpy's predictions misleading. One approach to this issue might be to monitor closely the changes in computed intent over time, letting the analyst adjudicate whether such observed changes are merely computational artifacts or in reality plausible enemy command behavior.

Turning to human factors, the requirement for the manual input of data by the operator may be detrimental to OpSpy's effectiveness, depending on the extent of the input required. In particular, the analyst has a challenging task in formulating a hypothesis, for he faces a complex tradeoff in choosing components of the hypothesis. On the output side, the analyst and his superiors will be required to review an enemy's predicted actions and decide whether they can understand and trust the tool's determinations. To make such judgments, humans will need insight into the underlying assumptions and the complex processes that generated the recommendations. Effective visualization, among other techniques, will be required to overcome this and other challenges common to most decision-support tools.

NO MERE TOY

Granted, the vision presented here is of much greater scope and ambition than what was actually researched and tested to date. However, many of OpSpy's scientifically challenging aspects had already been prototyped, and rigorous experiments have been conducted to confirm the technology's properties and capabilities.

The ability to determine the coefficients reflecting enemy intent, and to predict ECOA, has been demonstrated in simple, small-scale cases similar to the one described in Figure 4-2. Further, it has been proven that the algorithm can perform such tasks in time that is linearly proportional to the problem's size. It was also shown, in principle, that the same approach can be used to identify friendly probing actions that provide information about enemy intent. Much remains to be done, of course, but the key conceptual directions have been established with a degree of certainty.

Furthermore, technologies like OpSpy are not born in a vacuum. There is an extensive body of scientific work and technical applications of similar approaches used to identify unknown systems in complex, demanding applications such as industrial control and econometrics [4]. Although not entirely identical to military C^2 problems, such prior research and successful application lends some assurance of future success in military problems as well.

Overall, the implications of such technologies as OpSpy are staggering. For the first time in the history of warfare, military decision-makers can be offered a serious promise of systematic, theoretically rigorous, machine-assisted analysis of enemy intent and future actions. In current military C^2 practice, human analysts perform enemy intent identification without significant decision-support means. While automated tools do exist for war gaming [5], they do not attempt to identify enemy intent. Instead, they rely to some extent on preconceived enemy doctrine that is encoded in the rules of war gaming, and on elements of enemy intent identified and entered manually by human analysts. Unlike such tools, OpSpy offers a fundamentally different and novel capability to infer enemy intent directly from recently observed actions.

SUMMARY

Understanding enemy intent is a task that challenges even the most able commanders.

One way to represent enemy intent in a formal, rigorous manner is by a collection of coefficients that serve as a concise code, a kind of DNA of enemy behavior. Here, the coefficients reflect the enemy's apparent preferences, or values of goals, objectives, assets, and so on. Although an enemy most likely does not recognize such coefficients in his plans, they are the objective mathematical image of his actions.

To derive the code of enemy intent, one might use an analytical tool such as OpSpy described in this chapter. The input to the tool is the observed enemy action—preferably the most recent history of his actions, although more distant history can be useful as well. A database is used to maintain the records of such actions.

The OpSpy algorithm analyzes enemy choices—actions committed compared to the available alternatives—and constructs a mathematical system of inequalities reflecting such choices. Solving the system of inequalities, the algorithm derives a range of possible values for each coefficient.

These coefficients can then be used to predict the enemy's future behavior, his probable COAs. (A friendly COA is also required to compute the enemy COA.) Methods to conduct such predictions are discussed in later chapters.

Like in the scenario of this chapter, using a tool like OpSpy, an analyst can compute the code of intent from those enemy actions observed in the last several days of the campaign. He can then predict enemy COA and advise his commander that the friendly deployment plan might be far from optimal.

CHAPTER 5

HIGHLY DECENTRALIZED INTELLIGENCE AND C²

It is October, 2018 hours. A Blue force, consisting primarily of unmanned platforms—airborne, ground, and waterborne—is winning the fight. Confidently and systematically, swarms of Blue autonomic warriors find and destroy high-value Red targets while avoiding enemy defense hard points.

Competent and capable, the Red force responds by searching for the primary C² center. Destroying such a center, followed by similar attacks against fallback centers, should be an effective way to diminish Blue force effectiveness. Yet, strangely, the Blue force main C² center proves entirely elusive—the Red force fails to locate it.

With powerful electronic combat capabilities, the Red force's attention turns now to attempts to disrupt or spoof Blue communications. Equally strange, these communications seem to be entirely the hard-to-break, short-distance kind. In addition, the communications volume is very small, while the content does not seem to have any discernable meaning.

The Red forces undertake another approach—a rapid, unpredictable COA—where forces are shifted, and directions and tactics changed. It is supposed to work! By the time Blue intelligence delivers the news, by the time it gets exploited and delivered, by the time decisions accumulate, by the time detailed plans are worked out and percolate downward . . . Red's new approach should work, but it does not. The Blue forces seem to recognize these attempts almost instantaneously and take appropriate actions.

Surely, there must exist a super-powerful, super-intelligent mechanism that somehow enables Blue to accomplish this miracle of survivable, robust, precise, rapid C^2. Is it so? Is it possible to create a C^2 system in which communications delays between sensing and decision making become negligibly short? In which communications between C^2 nodes are virtually unnoticeable and extremely resistant to enemy interference? In which the destruction of one or multiple nodes has little or no impact on the overall system's functioning? Although such C^2 system properties may appear fantastic, one may find such a system literally underfoot. Ants use chemicals called pheromones to communicate information and make decisions in a very simple yet robust and efficient manner. Omnipresent, enormous armies of ants—some of the most successful and effective species on Earth—have been using C^2 systems with precisely such properties.

In our scenario, Blue succeeds because it uses a C^2 approach emulating the ants' pheromone-based method for C^2. Blue command defies the conventional vision of a huge underground installation with hundreds of staff officers peering into computer screens. Instead, Genies, a self-organized network of software agents residing on relatively small and simple devices or Hostility Observation and Sensing Terminals (HOSTs), which can carry Genies [1] and such other software agents as Avatars and Ghosts, guide the unmanned warriors' actions. Such devices are deployed in the battlespace to provide Blue forces with local, networked, and distributed military intelligence and control. While researchers at ERIM, Inc., headquartered in Ann Arbor, Michigan are exploring the concept and its corresponding technologies, the inspiration largely arises from a far more ancient and widely dispersed source [2].

THE POWER OF PHEROMONES

The Genie concept is inspired by the behavior of ants. Essentially leaderless, ants' elegantly coordinated work is controlled by their reaction to the smell of pheromones, chemical substances which ants produce and deposit on their surroundings. Here is a very simplified description of this control mechanism. Having found a desired object, such as a source of food, an ant continues its travels while squirting drops of a pheromone with a particular flavor that indicates "I found food." If an ant stumbled into a threat, it would squirt a different-flavored pheromone: "I found danger." In both cases, other ants smell the pheromone deposits and follow them either toward the food or away from danger.

Such pheromone deposits, made by multiple ants, gradually accumulate to create a field of pheromones. In every case, the environment provides three important functions to maintain the field: it aggregates deposits from multiple ants (data fusion), evaporates deposits over time to purge obsolete information (truth maintenance), and propagates pheromones from the deposit site to the nearby region (information dissemination). The pheromone deposits made by multiple ants, gradually accumulate to create a field of pheromones, of multiple flavors, a sophisticated and useful guide to the society of ants [3].

It is easy to draw an analogy between a forest populated by ants and a battlespace populated by autonomic warriors. Moreover, it is tempting to extend the ants' effective and robust control to that of autonomic warriors. But, what about the pheromones? Should we imagine our autonomic platforms swarming around the battlespace, all squirting and sniffing chemicals?

Probably not. Instead, let's try a different approach, using algorithmic—rather than chemical—pheromones. Unlike real pheromones, algorithmic pheromones are data structures kept in software agents' memories [4] (see Figure 5-1). However, similar to the real pheromones, autonomic platforms interpret the algorithmic pheromones as guidance toward targets and away from threats.

Although our analogy has just replaced chemicals with electrons, we still need a way to deposit the algorithmic pheromones throughout the real battlespace, for they need to be produced and used by real warriors. To do so, we would spread HOSTs, thousands of small hardware units containing a Global Positioning System (GPS), computer, radio communication package, and various sensors (for example, vibration, magnetometer, acoustic, infrared), in a somewhat random array throughout the area of interest, analogous to mine laying operations.

Inside each HOST is a software agent called a Genie. If HOST is the Genie's bottle, the Genie in turn is responsible for maintaining the pheromone information. As such, Genies continuously collect data on enemy forces through onboard sensors, and regularly share this information with neighboring Genies. In addition to onboard sources, Genies can accept and incorporate inputs from other sources, such as reports from satellites, UVs, and so on.

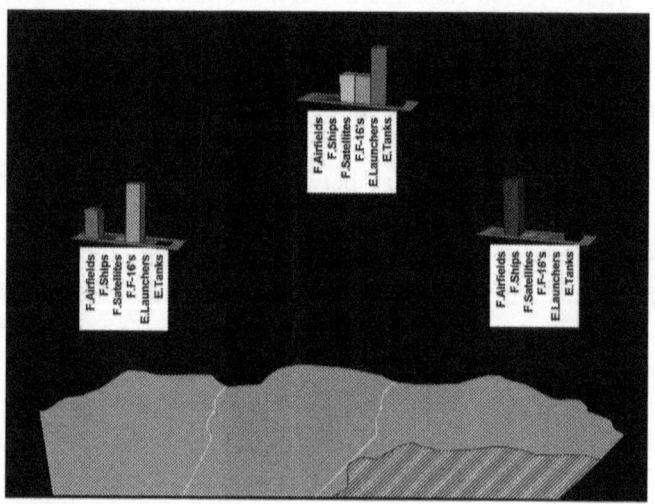

Figure 5-1. A Genie continuously maintains the picture of the situation through a collection of pheromone strengths corresponding to friendly and enemy assets in the area.

The pheromone field, used to guide friendly forces, is formed by the Genies' continuous information propagation and update throughout the battlespace. As friendly forces enter the area, they establish contact with local Genies, which provide them with continuous, real-time data on threats and targets. Genies also provide friendly warriors with guidance signals, which optimize routing toward targets and away from threats.

Although the Genie mechanism is inherently decentralized, that does not mean the commander cannot control it. Indeed, the Genies' computational processes can be modified dynamically by updating policies residing in the Genie itself—with policies being collections of rules reflecting the rules of engagement, doctrine, and commander's guidance.

The fundamental philosophy of the Genie approach is radical—a widely dispersed, decentralized, autonomous, largely self-organizing control of military forces. The potential advantages are numerous, for in spite of the pheromone model's inherent simplicity, the overall behavior that emerges in such a decentralized system can be very agile, robust, and adaptable, solving problems that are usually considered to require far more complex reasoning [5]. The concept lends itself well to increased autonomy and local decision-making. Units are not encumbered with the friction and inertia of traditional hierarchical C^2, and so can respond more rapidly to local changes in situation. Being an adaptive system with emergent behavior, such a C^2 system can degrade gracefully and self-heal under attack. Indeed, the operations can even continue when conventional C^2 nodes and channels are disrupted.

LIGHTS, CAMERA, ACTION

The Genie's ultimate value is to guide mobile warriors' actions, including a group of small, inexpensive UVs with limited onboard sensors [6]. For example, surveillance UVs would enter the battlespace, receiving and following guidance from Genies, which in turn use the pheromone field to direct each UV towards targets of interest. By monitoring reports of UVs traversing the battlespace, higher order C^2 nodes form an integrated picture of enemy dispositions, as shown in Figure 5-2.

Similarly, another operational concept uses UVs equipped with weapons to follow guidance from the pheromone field, precisely locating and destroying targets. Such combat UVs require only minimal onboard sensors to complete the strike, as all other essential tactical information derives from pheromone signals.

Clearly, it is but a short leap to extrapolate this technology for manned aircraft and other combat vehicles. One possible vision includes using a pheromone field to convey both enemy and friendly situations to conventional C² system decisions makers.

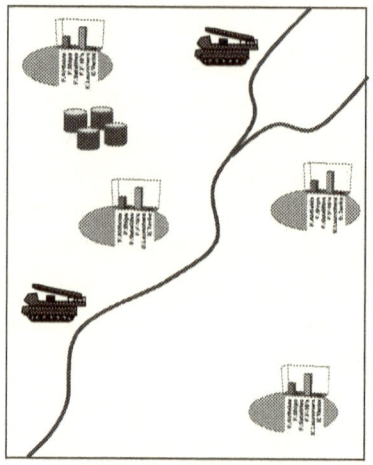
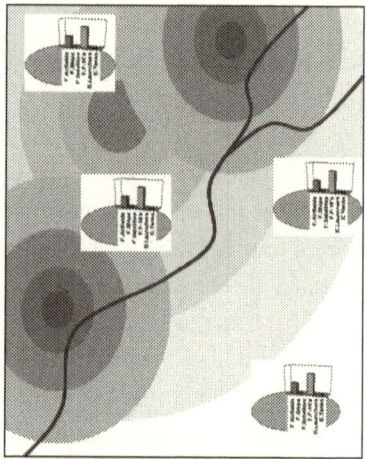

Physical Assets   **Field Representation**

Figure 5-2. Although an individual Genie acquires and maintains only the local situation, as a society these agents form a picture of the pheromone distribution fields throughout the battlespace.

Briefly, here is how it all works:

— Using techniques similar to those in mine laying, a large number of HOSTs containing Genies are deployed throughout the battlespace.
— Once a HOST lands, each Genie identifies neighboring Genies, establishes connections with them, and negotiates their respective geographic boundaries of responsibilities.

— Each Genie integrates local sensor information with externally supplied military intelligence indexed to its geographical region.
— The HOST is usually stationary, and its Genie is responsible for the surrounding area.
— Within its area of responsibility, each Genie uses onboard sensors to collect information regarding the enemy.
— Each Genie shares its information with other Genies, thereby forming more accurate awareness.
— UVs swarm over the battlespace, requiring guidance—when and where to move and how to act.
— The Genies provide information and guidance to the various mobile platforms entering the area.
— The mobile platforms execute their intended actions, such as surveillance or attack.

In some respects, the Genie's military intelligence function can be compared to the U.S. Air Force's Vietnam-era Igloo White Operation, in which large numbers of sensors were strewn along parts of the Ho Chi Minh Trail [7]. The sensors, disguised as plants and twigs, were able to detect the sounds of truck engines, vibrations, and even human scents. Collecting such signals, the sensors communicated them to Thailand-based central locations, which in turn used the data to direct bombings on the trail. While that operation's overall effects are debatable, the Genie differs from Igloo White in a fundamental way: using the pheromone mechanism, the Genie transcends being merely a collection device and becomes instead a distributed, decentralized C^2 system, able to guide autonomic warriors' local observations and actions.

THE ADVENTURES OF A SMART ROCK

October 19, 0500 hours, near the hamlet of Murilaka. An intelligence report arrives at the operations center of the 3rd Air Expeditionary Force: significant Red light ground forces are repositioning, moving through difficult terrain that offers excellent opportunities for cover and concealment, including extensive vegetation and a rural population friendly to Red.

October 19, AM, 5ᵗʰ Division Supply Base, port Dikturi. Within four hours, using automated factory-in-a-container at the local supply base, thousands of HOST devices are prepared and mission-configured. Each one is different: rocks, clumps of dirt, tree branches, and so on. Besides being automatically painted in colors most suitable to the local soil and vegetation, the HOSTs are also programmed to fine-tune and allocate sensor and computational resources to the most likely types of forces, terrain, and weather conditions.

October 19, AM/PM. During the night, airdrops and long-range artillery deliver the HOSTs to the Red force's most likely routes and locations.

October 19, 1300 hours, vicinity of Gartala Mountain. A HOST disguised as a rock falls from the sky and hits the wooded slope. Bouncing and rolling down, the rock, wedged between huge boulders, finally stops. Having determined its location—the narrow point of a mountain creek known to locals as Bugarga—the rock's Genie issues a short-range radio broadcast: "I am Genie-D8J5W1 at location LG2345T6789. Are there any friends nearby?" A few seconds later, it receives and acknowledges responses from several nearby Genies. Now knowing the identity and location of its five nearest neighbors—those within range of its communications equipment—the Genie has established its own communications network.

Clearly, then, Genies automatically and autonomously organize and configure their entire society. Here, this Genie joins the greater network of Genies, all of which can propagate information to the entire group through the combined relay of their immediate neighbors. In addition, a Genie automatically identifies and negotiates its area of responsibility.

October 21, 0300 hours, Gartala Mountain. Following along the swift mountain creek Bugarga, a platoon of Red infantry marches with lightweight yet highly lethal modern weaponry, including surface-to-surface missiles containing biological warheads, smart surface-to-air missiles, intelligent mines, radars, sensors, and IW equipment. Wedged between two boulders, and half-immersed in the creek, a fist-sized rock, virtually indistinguishable from thousands of nearby rocks, watches the passing Reds with its multiple eyes and ears—optical, infrared, olfactory, acoustic, magnetic. In addition, the Genie receives input from the HOST's onboard sensors, interprets them as pheromone signals, and compares them with pheromone signals communicated by neighbors. In human terms, the rock thinks something like, "Here they are. Finally. There was nothing for 37 hours except for a couple of wild pigs. Now their combined signature matches Type 46D, Pedestrian Humans Laden with Hi-Tech Equipment. That's consistent with what I heard 73 minutes ago from my neighbor, Genie-F2T8Q9. It is 3.7 miles to the SE of me, and also reported between 10 and 100 personnel of Type46D. The speed of movement is within bounds for Type46D. It all matches."

The Genie immediately begins issuing brief bursts of short-range radio communications, propagating a Type46D pheromone to its neighboring Genies, notifying them of detecting a force of Type46D, approximately 10 to 100 in strength. While the actual message contains only a sender ID, a pheromone type indicator, and pheromone strength, this simple information is exactly what the fellow Genies need to do their work—but much less comprehensible or useful to anyone who might manage to intercept it. Here, the Genie's neighbors pick up its message, and, based on this new information, adjust their perceptions of the Red assets. The overall picture emerges bottom-up, via propagation within the network of Genies.

Three hours later, our smart rock and its neighbors downgrade the strength of pheromone Type46D at this location by a factor of three. The rock will repeat this downgrading again and again as time passes. Why? Although the Genie may not know the detachment's direction or speed of movement, the lack of any subsequent sensory or military intelligence reports reinforcing initial contact suggests that the Red force is probably moving away. As hours pass, the Genie periodically reduces the strength of its recently detected subjects. While this evaporation of evidence process is not necessarily accurate, it is a robust way to purge information as it becomes less recent and more doubtful.

October 21, 0700 hours, Gartala Mountain. The rock receives a call from a Blue UGV—a small, uninhabited ground vehicle about a mile north, collecting intelligence, crawling in the bushes along the crest of a hill. After establishing a proper electronic handshake, the two exchange what they know. The UGV tells the rock that about 20 miles to the north there is a convoy of all-terrain supply vehicles. Although it is outside the area of interest of either the rock or its immediate neighbors, the rock takes the responsibility for conveying the information to the Genies responsible for that area. This is not a difficult task, for the rock immediately broadcasts the info to its neighbors, which in turn propagate the info through the network until it reaches the right agents. As such, the Genies' network is an effective mechanism for getting information from any source to the right consumer, even when neither provider nor consumer are aware of each other.

Throughout this time, at irregular but frequent intervals, the rock receives messages from its neighbors. Like it, they observe the battlespace, make conclusions, and share them. At one point, a neighbor located 1.5 miles SW signals, "I have a weak observation of a group of pedestrian humans, numbering five to 10, no specific equipment identified, type uncertain." Hearing this, our rock is not entirely trusting: what if this message is some sort of enemy spoof? Having compared this information with its own data, as well as with information recently received from other neighbors, the rock judiciously updates its database.

October 21, 1200 hours, Murilaka. Some 200 miles away from our smart rock, a Red Special Forces unit has been in hiding for the last four days, waiting near the Blue operations center, adjacent to the small hamlet of Murilaka. Concerned about heavy, accurate, and effective Blue attacks, Red commanders issue orders to execute preplanned attacks against Blue C^2 centers. After receiving the signal, and using high-tech penetration weapons to breech the central C^2 module interior, the Red unit executes a deadly effective suicide attack. Another unit executes a simultaneous attack against the secondary operations center, which turns out to be less effective but is still strong enough to incapacitate this center for more than 48 hours. At a well-calculated interval, a competent and wide-ranging IW attack plays even greater havoc with Blue efforts to restore its C^2 capabilities. The rock, meanwhile, is entirely unaffected by these sad events, for none of his superiors' misfortunes prevents it from performing its duties. Indeed, the society of Genies can be controlled from a central C^2 node, but the society is not dependent on such centralized control. As in the current example, the destruction of a central node does not incapacitate the network of agents.

October 21, 1600 hours, Gartala Mountain. The rock detects a call: an UV on a seek-and-destroy mission has just entered the airspace over the rock. After the handshake, the UV explains both its mission and constraints: its payload is best suited for hunting the Reds' small supply vehicles, but the UV itself is vulnerable to SA-X man-portable surface-to-air missiles and should avoid those. Where should the UV go now: any suggestions? A furious exchange of short-range communication bursts ensues within the Genie community. While no individual Genie is capable of making a useful suggestion to the visiting UV, together, using a technique called Ghost Agents, they formulate a detailed mission plan, complete with specific targets, threats, and a proposed routing that will optimize effectiveness against the target vehicles while avoiding the SA-X envelopes. In all, the process takes less than one minute, from the UV's initial contact until the rock conveys the recommended mission plan. In the last 24 hours, it is already the 17th such encounter with itinerant, unmanned warriors for this rock alone. Here, interactions of multiple agents—Genies, Avatars, and Ghosts—enable agile, decentralized C² of autonomic warriors.

October 21, 2300 hours, Gartala Mountain. One all-terrain vehicle and seven Red soldiers are straggling along the creek, all that's left of the supply convoy destroyed five hours earlier by several cooperating Blue UVs staging a well-executed attack. If our rock could feel pride, it would be pleased to know that the leading UV was the one it advised a few hours earlier, and that the attack followed the mission plan it helped to generate. However, being just an unfeeling collection of electronics encased into a plastic shell, the smart rock merely continues to do its job, observing the supply column remnants, computing appropriate pheromone numbers, updating its database, and issuing messages to its neighbors.

BODY AND SOUL

Yes, but how do Genies actually work? First, let's look at the messages Genies receive and process. While onboard sensors provide much of the information, messages from neighboring Genies are also a key information source.

Each HOST contains a small hardware unit containing, among other modules, several onboard sensors (for example, vibration, magnetometer, acoustic, infrared) that collect data on the enemy. Since this sensor data is pre-processed onboard, by the time it enters the Genie it is expressed as messages in the pheromone language, something like "pheromone #534, 1.73." Translated into English, such a message might read as follows: "I am sensing what appears to be Red Light Armor, approximately company-sized."

Much could be said, and argued, about sensors—types, proper combinations, range, data fusion—as well as about the tradeoff between the cost of HOST hardware and the precision of identification, and so on. However, such issues are best left for a separate study—as is the important topic of Automatic Target Recognition, and how and to what extent Genies would perform this difficult task. Indeed, an extensive body of past and current research deals with this challenge.

In any case, it is understood that the accuracy of sensory observation and identification will not be perfect, but the distributed, collaborative nature of pheromone thinking will help in this respect. By combining and mutually reinforcing the most common signals, the society of Genies will average out a significant part of errors. Similarly, because localized attempts at deception will largely be filtered out by cooperating Genies, resistance to enemy deception is also likely to be higher.

In addition to onboard sources, the Genie can accept and incorporate input from external sources, such as satellites, UVs, and so on. While the nature of such messages is similar to those discussed above, there is an important difference: while observing a certain asset in a certain location, non-HOST observers will not necessarily be able to communicate directly to the HOST responsible for that location. Indeed, at that time these non-HOST observers would not even know which HOST might be responsible for that location. Therefore, the non-HOST observers will communicate to whatever HOST happens to be reachable, and the message will also include the request to pass down the information to whichever HOST is responsible for the area specified.

OVER THE BACK FENCE

Besides dealing with data from onboard and off board sensors, a Genie busily communicates with other Genies. The most common and frequent message exchanges are those by which Genies propagate pheromones throughout their network. Examples of elements that might be included in such messages expressing Genie pheromone parameters include:

— Time stamp—the time when the pheromone message is generated
— Pheromone source—the location or ID of the unit generating the message
— Pheromone class—the type of the asset and/or activity the Genie can observe
— Pheromone strength—the amount of the given class' pheromone that the Genie is advising its neighbors to add to their stored values
— Uncertainty factor—the certainty of the source about message content
— Evaporation factor—the number reflecting how rapidly the neighboring Genies should depreciate the information (that is, decrease the amount of a stored pheromone); a message originator's recommendation to its neighbors
— Propagation factor—the number reflecting how aggressively the neighboring Genies should propagate information further through the network; a message originator's recommendation to its neighbors
— Update frequency—the frequency with which the pheromone should be updated for evaporation and propagation.

Would a Genie be overwhelmed by all these communications? How numerous, frequent, and diverse will the message flow be arriving to a Genie? One can envision certain order-of-magnitude numbers. Assuming one pheromone for each significant type of a military asset, a typical Genie will have a repertoire of roughly 100 pheromones—but only a few of them would be relevant at any given time. In a single communications burst, perhaps as many as 10 pheromones will be updated. The bursts, depending on the OPTEMPO, the agility of enemy assets, and the update frequency of associated pheromones, might be issued in periods between 10 seconds and a few minutes. In addition, the number of neighbors with whom a Genie exchanges messages will be three to five, with the amount of incoming messages increasing correspondingly.

Although less frequently used, other types of messages from neighboring Genies will also be important, including messages:

— Introducing new neighbors to each other
— Informing about a neighbor's status (for example, whether it is still alive)
— Asking to propagate information arriving from non-HOST observers

HOW A GENIE THINKS

Having received all these messages, a Genie must think about their meaning and make useful conclusions. Yet the computations performed by each individual Genie are relatively simple: as in other highly distributed systems, the sophistication, agility, and robustness of the resulting intelligent behavior emerges not from the power of any individual entity, but instead from the synergistic collaboration of multiple simple entities.

A Genie perceives and internally represents the world as a collection of computational pheromones. Therefore, a Genie's mind assigns a pheromone (essentially a scalar variable, for example, a number like .175) to each significant type of friendly and enemy asset. To avoid an excessive number of pheromones, assets are viewed at a reasonable level of abstraction. For example, if onboard sensors cannot distinguish between different types of armored assets, there is no point in programming the Genie with codes for different types of tanks and armored personnel carriers. Instead, a single pheromone will be dedicated to armored assets in general.

For each pheromone, the Genie keeps a number reflecting its amount, that is, the cumulative, time-adjusted asset strength the Genie has detected in its area of responsibility. Cumulative, here, indicates that the Genie continually adds new signals about this pheromone, whether they arrive from onboard sensors or from such other sources as neighboring Genies. Time-adjusted indicates that the Genie also subtracts from the strength of the pheromone as time passes.

Suppose a Genie was programmed to consider Heavy Armor as pheromone #25, Infantry as pheromone #73, and Radar as pheromone #137. The Genie, deployed in a certain battlespace location, has been at work for some time. Looking into its memory, we see 25-0.03; 73-2.35; 137-1.89. Roughly, the Genie says that it has very little Heavy Armor pheromone, meaning that it either did not sense much of it, or sensed it a long time ago. Instead, there is a significant amount of Infantry pheromone, and of Radars. Therefore, the Genie must have received extensive signals about these assets, and relatively recently.

Of course, this is just a snapshot of the Genie's memory for a given time moment. Because the pheromone picture is completely dynamic, changing with time and propagating through space, if we were to look into the Genie's memory 10 minutes later we might see a far different picture.

AGGREGATION, PROPAGATION, EVAPORATION

Even the best and most accurate sensory data is of little value unless it is used in a fusion from multiple sources, in communication and cross-cueing, and in continuous verification and maintenance. When such a processing is applied, even relatively inaccurate and limited data can be enhanced in value. Thus, in the world depicted by pheromones, Genies continuously update their view by three processes:

— Aggregation of pheromone strengths from multiple sources (fusion)
— Propagation of pheromone strength to neighboring Genies (communication)
— Evaporation of pheromone strength over time (truth maintenance)

Aggregation

Suppose a Genie has a record in its memory regarding pheromone #25 (Heavy Armor): 25—.03. Then the Genie receives a message from its onboard sensors, "I sense pheromone #25 at a .75 intensity." At this point, the Genie adds the new strength to the one it has in memory, and the new record looks like this: 25—.78. Actually, the process is more complicated, for the amount of additional strength may not be .75, but instead may be adjusted based on time elapsed since the previous message, reliability of the sensor collection, and so on. Nevertheless, the gist is that simple.

Other sources, particularly non-HOST military intelligence and neighboring Genies, can also send similar messages, which are treated via similar aggregation. The value of this process is that it fuses information from various sources, and reinforces the information if signals continue to arrive over time and from multiple sources. For example, when multiple ants deposit the pheromone that indicates "I found food," it gradually aggregates into useful paths that lead to the food sources. However, aggregation is a mechanism that does not work in isolation; other processes such as propagation and evaporation must support it.

Propagation

Just as ants' pheromones are carried to neighboring locations by wind and water, so are the computational pheromones propagated through the network of Genies. Periodically, each Genie takes a fraction of each pheromone's strength it has in its memory and sends it to neighboring Genies in a message, "Add strength .06 to your pheromone #25." The neighboring Genies receive this message and do as instructed—they increment pheromone #25 strength by the suggested amount. The exact details regarding periodicity, fractions propagated, and how much neighbors actually add to their records are far more complicated. Nevertheless, there is continuous information propagation through the entire network of Genies.

Such a mechanism serves multiple purposes. First, it states that the asset's location, as sensed by a particular Genie, may not be precise. Second, it reflects that assets are continuously on the move, changing their locations over time. Third, it helps to smooth out the pattern of asset distribution, even if one of the Genies is unable to detect assets in its own area. Fourth, it allows Genies to compare notes, either reinforcing mutually supportive observations or canceling spurious and deceptive signals.

Evaporation

For the purposes of maintaining true information and filtering out false information, the evaporation process is even more important than propagation. The Genie performs evaporation as follows: suppose, at 1600 hours, that a Genie has in its memory a record regarding pheromone #25 (Heavy Armor), 25-.78. An hour passes. At 1700 hours, the Genie reviews the record, reduces the number in half, and produces a new number, 25-.39. In essence, the Genie is saying, "Look, this information is now aged. It is an hour old, and I cannot give it the same credence as I did an hour ago." Of course, these numbers are simply an example; the actual period of aging, and the amount of reduction over time, are dependent on many factors—the nature of the asset represented by the pheromone, the environment, the OPTEMPO, and so on.

This evaporation mechanism is simple, inexpensive, yet powerful, for it allows the system to forget gracefully information that is not reinforced, and to remember information, which is reinforced. One may compare this simple mechanism with sophisticated and complex computational algorithms used to maintain truth in technologies that use logic-based computations. Here, the evaporation mechanism is dramatically simpler and less expensive, yet allows the Genie to deal with such difficult issues as outdated truth, half-truth, and uncertain truth.

MOBILE WARRIORS

In our scenario, in addition to HOSTs, mobile, active platforms, such as UVs, populate the battlespace. Another of the Genies' key roles is to assist these itinerant warriors, beginning when a mobile platform arrives into a Genie's area of responsibility and establishes contact. Once the platform registers with the Genie, the Genie then creates a new agent called Avatar, which like the Genie resides on the HOST. This Avatar, an image of the particular mobile warrior, is its electronic ambassador within the HOST. Unlike the Genie, however, the Avatar resides on this HOST only temporarily. As soon as the mobile warrior moves to the area of another HOST, the Avatar moves to that other HOST as well.

When the Avatar, acting as its mobile warrior's representative, needs information or guidance on how to continue the warrior's mission, the Avatar issues a request to the Genie (see Figure 5-3). Several types of requests are possible, depending on the warrior's mission, status, and internal capabilities, including:

— "Tell me what assets are in location L."
— "Tell me the strength of pheromone X in location L."
— "Tell me where I should go if I am looking for targets corresponding to pheromone X and want to avoid threats of type Z."

Depending on platform type, mission, and guidance, the Avatar may or may not follow the Genie's recommendations. In any case, the Avatar communicates guidance decisions to the platform.

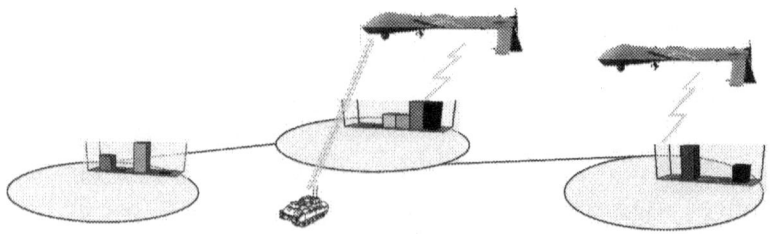

Figure 5-3. Using the pheromone field and Ghosts, Genies recommend a path to Mobile Warriors, where Mobile Warriors find the general target area and then home in on targets using on-board means.

Having received a request of the type, "Tell me what assets are in location L," the Genie responsible for location L generates a response of the type, "In my area, there is a Red SAM-XX Battery, a Red fuel dump, a Blue special operations forces unit, and so on." A message of the type, "Tell me strength of pheromone X in location L," is merely a simplified version of the same request and receives a corresponding report.

A related service that a Genie can offer to an Avatar is to suggest in which direction a warrior can go to find stronger pheromones of a given type. To provide this service, each Genie needs to be able to compute the local gradient of a specified pheromone. Querying its immediate neighbors, the Genie can determine which offers the strongest pheromone concentration.

A request such as, "Tell me where I should go if I am looking for targets corresponding to pheromone X and want to avoid threats of type Z," is significantly more complex. To respond, the Genie must perform different, more sophisticated computations using Ghosts, to be discussed shortly. Ultimately, the response will be of the type, "Head to the way point with latitude X and longitude Y," or possibly a more elaborate definition of the movement path.

The greater the number, and the more complex the requests, the heavier is the Genie workload. Indeed, the number and frequency of mobile warrior requests arriving to a given Genie can be significant, and will depend on platform density in the battlespace, broadcast range, and call frequency from a given UV, possibly on the order of 30 seconds. When many mobile platforms enter a given Genie's area, a number of them will undoubtedly have the same or very similar requests. The Genie may take advantage of this fact by storing its most recent request-response pairs, then using them to provide answers to the mobile platforms without the expense of recomputation.

GHOST AGENTS

For a Genie to answer a warrior's question like, "Tell me what path I should follow if I am looking for targets corresponding to pheromone X and want to avoid threats of type Z," it must explore the pheromone field. That search can be quite complex, effectively requiring some degree of stochasticity, that is, judicious random explorations. That's where Ghost Agents enter the picture, for they are the ones to which the Genies delegate the stochastic aspect of the search. Each Ghost is an electronic image of the mobile warrior, and each rapidly explores the pheromone field by moving from Genie to Genie. Based on the results of multiple Ghost explorations, a Genie suggests the warrior's best path.

Like Avatars, each Ghost represents a particular mobile platform. However, Avatars and Ghosts differ in several ways:

— While there is only one Avatar for each platform, there may be many Ghosts, all exploring different possible behaviors.
— While the Avatar makes decisions for the platform and issues commands to it, the Ghosts do not interact directly with the platform, instead exploring and summarizing the pheromone field for the Avatar.
— While the Avatar is active as long as the platform is on its mission, Ghosts have much shorter lifetimes.

Here is how Ghosts, Avatars, and Genies work together to find and suggest a good path for a mobile warrior, as is shown in Figure 5-4. On a periodic basis, the Avatar requests that the Genie send out a set of Ghosts, with the periodicity very short compared to the time it takes the Avatar to move from one Genie to the next. The Avatar, which must move physically from the vicinity of one HOST to another, is constrained by the physics of its platform. Ghosts, however, move between Genies at the speed of light, pausing only to consult their current Genie for guidance. An Avatar might launch 10 Ghosts every second and specify that each Ghost has a lifetime of 50 seconds. Thus, at any moment, there might be some 500 Ghosts active for each Avatar.

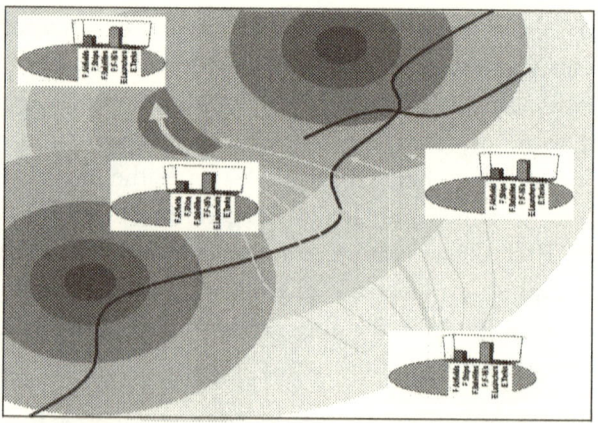

Figure 5-4. A Genie sends out waves of Ghost Agents, which sample the pheromone field and deposit their own track pheromone. The track pheromone then condenses into an optimal path through the pheromone landscape.

Launched by an Avatar, the Ghosts search the pheromone field by moving from one Genie to another stochastically. As a result of the Ghosts' interactions with each other, the pheromones they deposit form a narrow ridge that indicates the best path for the platform to follow. The Avatar then follows this path, moving from Genie to Genie, all the while continually sending out new waves of Ghosts to update and modify the portion of the path not yet followed. Thus, the overall process is the continuous interaction of Genies (responsible for aggregating, propagating, and evaporating pheromones), Ghosts (responsible for stochastically searching the pheromone field), and Avatars (responsible for following the initial portion of the path, and thus changing the starting point from which subsequent generations of its Ghosts search the pheromone field and update its path).

DISTRIBUTED IN SPACE AND TIME

Agents like Genies, Avatars, and Ghosts comprise a naturally distributed architecture, with its computational modules, or agents, independent computational processes. The three agents' interactions are, by design, strongly local—that is, at any given time one agent only interacts with one or a few other agents, and when this set of acquaintances changes, it does so in a bounded way. For example, interactions among Genies are limited to those agents that represent adjacent physical regions. Each Avatar interacts only with one Genie at a time; when the Avatar moves to another Genie, the Avatar can go only to one adjacent. Therefore, because of the localized nature of interactions, the society of such agents lends itself to distribution across multiple, networked computers.

The agents' society dynamics emerge in a bottom-up fashion through the interactions of its various distributed processes. Indeed, the way the agents behave in time is not controlled but rises through the agents' own interactions [8].

A very practical concern is that spatial distribution depends on the deployment phase, which in itself could be challenging. HOSTs could be deployed throughout the physical battlespace by means analogous to mine laying (for example, artillery, air drops), or via small UVs. For example, the Gator mine deployment system has long range and allows rapid minefield emplacement anywhere that can be reached by tactical aircraft. Similar means can be employed for distributing HOSTs in the hostile battlespace.

As with mine warfare, however, one of the potential shortcomings intrinsic in deploying the devices is that deployment itself may be high-risk (to the delivery vehicles). Given that the HOST array could well be deployed in enemy territory, the enemy will likely defend that space. The requirement to create an array over the whole space to be covered, not just around certain point or area targets, also complicates the delivery problem. Therefore, delivery vehicles will not necessarily have the luxury of avoiding threats along the way. In addition, aircraft mine laying has traditionally required a rather benign, non-maneuvering flight profile, potentially worrisome against a sophisticated air defense system. Similarly, given artillery's limited range, the deploying force will need to be in relatively close proximity to enemy territory. To cover the array on the far side of the battlespace, the delivery vehicle may well have to be inside enemy territory. Clearly, then, a careful analysis of the deployment phase is required, possibly considering alternative methods and technologies.

HUMANS AND ANTS

Because the system's state is spread out, it is hard for users to understand the causal link between a command decision and system outcome. Further, it is not clear what dial to turn to achieve a desired result. In addition, because the interactions among elements are nonlinear, bottom-up emergent architectures can produce system behavior that is qualitatively more complex than individual elements' behavior. Such nonlinearity means that system dynamics can become formally chaotic, leading to divergent trajectories for similar initial conditions, making even deterministic behavior formally unpredictable over any but very short time periods. (For that matter, unpredictability can appear even when the individual steps are completely deterministic.) Unpredictability and non-determinism mean that even if people could grasp the overall state of the system, they still couldn't predict with certainty what it would do.

One approach to enable human understanding of the system is visualization, and in fact the overall Genie architecture naturally lends itself to powerful visualization techniques. Distribution, meaning that the system is highly modular, together with simple computations, enables a variety of visualization interfaces, some of which have been demonstrated by the Genie research team. Even though detailed system behavior trajectories may be unpredictable in the chaotic regime, a higher-level structure can be detected, visualized, and used to manage the system [9].

The Genie approach's fundamental military philosophy—widely dispersed, decentralized, autonomous control of military forces— may also be problematic in some doctrinal circles. To take maximum advantage of this approach would require much more autonomy for the individual warfighter than is currently permitted. Since World War II, the U.S. Armed Forces have been moving toward a more centralized vision of C^2, as embodied in the Joint Chiefs of Staff and its formal doctrine and structure. This centralization has also been aided by the rapid evolution of high technology, which allows commanders the direct capability to control battlefield events oceans away. Yet, many tacticians would argue that some efficiency has been lost in doing so. Furthermore, in the last half century an active enemy has not seriously threatened intercontinental connectivity, so the robustness of such a centralized system may be suspect. Indeed, a current move in doctrinal thinking is toward more decentralized C^2 architectures that imply a willingness on the part of commanders to consider alternative doctrine [10].

AGILE, ADAPTABLE, VERSATILE

Alternative, decentralized doctrine may offer significant military advantages. Local decision-making enables individual units to react more quickly to local conditions. Such unit autonomy has at least two beneficial results: when conventional C^2 channels are disrupted, units can continue operating; and when the friction and inertia of hierarchical C² does not encumber them, units can respond more rapidly to local changes.

Because Genies and other agents interact directly only in a local region, the system can be scaled up without increasing the load on any existing agents or their processors. Emergent dynamics enables scaling to take place in real time, since newly added units will integrate themselves into the system as it runs.

The general nature of the underlying model means that a Genie-based system offers multi-mission, multi-task, and multi-echelon applicability. Indeed, this system can be used in a wide range of missions (for example, ground and naval as well as air units, manned or unmanned), addressing multiple tasks (mission planning, asset allocation, mission control, training support as a smart OpFor) at different command echelons.

Unlike conventional computation, which is all-or-nothing, a Genie-based system has the properties of any-time computation. On the one hand, conventional computation requires a certain length of time and produces a single answer. If this time is not available, however, no answer can be given; if more time is available, that extra time does not enhance the answer. On the other hand, emergent algorithms, like those used in Genies, are any-time computations, producing an approximate result very quickly, and then refining it as time passes. Therefore, even when time is extremely limited any-time computations give some benefit. As computation continues, any-time algorithms give additional benefit, enabling commanders to trade off decision time against decision quality.

The system's spatial distribution, as well as its components' temporal dynamics, yields any-time performance results. For example, as developed by Ghosts, the pheromone path forms quickly—but widely and crookedly. As time elapses, the path becomes narrower and straighter, giving better guidance to the Blue force. Because each component has only limited, local connections to the rest of the system, and because the system is simple, it completes processing quickly and yields initial results. As local computations' results propagate to more distant elements, local inconsistencies diminish, yielding an improved overall picture.

Further, because emergent systems organize themselves from the bottom up, they can self-configure with minimal top-down direction. Such a capability also enables these systems to be deployed in hostile environments, where detailed supervision by friendly forces cannot be guaranteed. In addition, the ability to self-configure contributes toward graceful degradation, enabling systems to reconfigure when they are disrupted.

Indeed, a direct relation exists between the non-determinism and unpredictability of emergent architectures and their ability to configure and heal themselves. Non-determinism is a powerful mechanism for searching out alternatives when an initial plan fails. Unpredictability manifests the superimposition of many accessible system trajectories in the chaotic regime. With appropriate techniques, the system can select among these when reconfiguration is needed. There is, in fact, evidence that any system capable of self-organization must operate on the border of the chaotic regime, since only in this regime are multiple trajectories simultaneously accessible.

FIT TO SURVIVE

Highly distributed systems—from ants to bees, from Internet to market economies—tend to be highly survivable. Killing a few bees will not stop the swarm's attack, and failure of a few Internet links will not stop e-mail general delivery. Similarly, the society of Genies is highly survivable: destroying some HOSTs, or jamming some links, will not hobble the entire system.

Although one might assume that a Genie network requires exceptionally robust connectivity, in fact, inter-HOST communication does not need to be any more robust than that of other systems. Furthermore, the ability of the HOST network to configure itself, and to heal itself when disrupted, suggests that it can handle communication faults more gracefully than other systems. Still, radio communications in a hostile environment can be extremely difficult. Therefore, the ability of many HOSTs inside enemy territory to maintain continuous radio contact over an extended period is an issue that will require careful consideration. Furthermore, enemy technology will eventually be able to develop receivers capable of breaking through countermeasures, picking up HOST transmissions, and either disrupting them or triangulating their positions to capture the hardware.

Although the ability of the enemy to disrupt, destroy, or capture HOSTs should not be underestimated, distribution, as well as the system's emergent dynamics, enables it to degrade gracefully as components are removed or disabled. Distribution here means that the computational load is shared across many processors, so that the removal of one results in only a fractional loss of the overall computational resource. Emergent dynamics enables the remaining processors to redistribute the particular functions that have been lost.

Both spatial and temporal features give the society of agents important security benefits. Because communications are local, they can be implemented with low-power transmitters that are less susceptible to tracking and interception. (Indeed, there is no need for the signal to persist over long distances.) Temporal features that contribute to security include non-determinism, the underlying algorithms' simplicity, and the information the algorithms use. Although non-determinism hinders our ability to understand and optimize our own system, it also hinders the adversary's ability to understand what we are doing. In addition, there are two security implications regarding the simplicity of the information our agents exchange. First, because there is little explicit semantic content in pheromone information, it is meaningful only in the context of the entire system; therefore, an adversary would have to model the entire system to understand the significance of any single intercepted message. Second, because the information needed in any single interchange is very limited, it can be transmitted in compact bursts, which are more difficult to intercept than are sustained transmissions.

SUMMARY

Using pheromones—chemicals that ants squirt on the ground under various conditions—ants and other insects are capable of controlling complex actions of their large societies. Accumulated signatures of these chemicals help the ants to make decisions. Although a very simple mechanism, it leads to very sophisticated emergent behavior.

Loosely based on this analogy, a society of Genies constitutes a radically distributed mechanism for collecting intelligence and making distributed, emerging decisions about military actions. Each Genie resides on a small computing and communicating device—disguised as a rock, twig, and so on. Dispersed in large quantities over the hostile battlespace, Genies use on-board sensors to collect information about surrounding enemy and friendly forces. Internally, Genies represent this information as a collection of numbers that they handle similarly to pheromones—they computationally aggregate this numbers, evaporate them, and propagate from one Genie to another. From these interactions, a shared picture of the overall situation emerges.

Genies also guide such autonomous warriors as UCAVs. Using the accumulated pheromonic picture of situation, Genies and other computational agents collaboratively develop mission plans for the itinerant warriors, adjusting them rapidly as the situation changes. Thus, Genies are a C^2 mechanism.

A system based on Genies may be difficult for a human to understand and control, and may not be consistent with conventional military doctrines. However, a Genie-based system could be very robust against local disruptions, could adjust to new events in an agile manner, and self-heal even when partially destroyed.

PART III
OPERATIONS PLANNING

CHAPTER 6

AGILE PLANNING
AND SCHEDULING

On a clear, cool February day in 1983, nine specially chosen individuals, representing each of the military services and defense transportation agencies, gathered at the Headquarters of the Joint Deployment Agency in Tampa, Florida. Their task was to shape the future of military planning and execution.

Planning and execution were considered separate and distinct disciplines in the military, separated organizationally and technologically. The Joint Operations Planning System (JOPS) was a tool for deliberate planning. Truckloads of plans were developed, each refined to squeeze every bit of efficiency from the military transportation system. Computer printouts listing every detail of imaginary missions filled cavernous rooms. Planners studied these printouts looking for even greater efficiencies.

Unfortunately, operators used different systems to support execution of such deliberate plans. Even though military planners documented every assumption, constraint and restraint for the planned operation, it was impossible to trace the impact of these factors on individual actions. The scheduling systems could factor in all of these items, as they created the massive schedules, but the systems lacked the agility to respond to simple changes—like a winter snowstorm in the Midwest or spring floods in Georgia.

The systems sought to evaluate the overall feasibility of the entire plan. Even minor changes required that the entire plan be set-up and rerun. Execution systems focused on immediate requirements, typically without regard for the overall plan. The massive, highly efficient plans couldn't be incrementally adjusted—and therefore were abandoned.

The individuals who came together that cold winter day created a vision for the Joint Operations Planning and Execution System (JOPES). Their vision called for a new generation of technologies that would assist human decision-makers. First, they demanded an end to the artificial division between planning and execution, calling instead for the two to be a single process. Second, they called for an expansion of technological support to include employment and logistics aspects of a total planning and execution system. Unfortunately, that vision was doomed: technology, they were told, could not support the capabilities requested.

IS THIS ACTUAL CLOCK TIME?

Today, a solution for uniting planning and execution may be closer to reality. Several challenges (data integration, multi-level security, knowledge acquisition, organizational structures) remain before a truly seamless, global planning and execution system is ready for use, but technology is no longer delaying this inevitable advance. For example, a joint effort of SRI International Artificial Intelligence Center, Menlo Park, California, and Carnegie Mellon University, Pittsburgh, Pennsylvania, offers an approach toward solving the issues that separated military planning and execution for almost a half-century. Their solution, called JPS, for Joint Planner-Scheduler, implements an incremental, tightly coupled approach to a planning and execution system.

Several concepts mentioned in previous and subsequent paragraphs need to be defined. The first concept involves the distinction between planning, scheduling, and execution. Planning focuses on identifying high-level objectives and then decomposing them into a set of time-sequenced actions. In essence, planning identifies what tasks need to be performed and when (in a relative sense) the tasks should be completed.

Scheduling focuses on time in the sense of actual clock time and what resources should be assigned to accomplish each task. At the operational level of command, where the Joint Planner-Scheduler is designed to operate, execution involves identifying the tasks to be performed, allocating resources to the tasks, establishing the time and sequence for task accomplishment, and assessing results. The process is repeated following the assessment, with the added consideration of the inertia associated with previously assigned tasks and the cost of reallocation.

HOW INTENSE IS INTENSE?

The second concept requiring definition is the concept of intensity, which will be used frequently in the following discussion. Intensity provides a common measurement of both resource requirements and capabilities. When applied to an individual task, task intensity reflects the allocation of tasks necessary to satisfy a specific objective. Intensity can also be used to represent a plan's total resource requirements over time by measuring the intensity requirements of all plan objectives.

Task intensity relates to the objective and reflects the level of resource usage necessary to achieve that objective. For example, the objective to gain and maintain air superiority could call for attacks against a specific airfield. One can use various strategies to eliminate the enemy's ability to use the airfield. Attacks can be directed against the runway, fuel systems, control facilities, munitions, aircraft parking shelters, or any combination of the above. Each of these actions calls for factors calculating the intensity of the attack such as, number of sorties, or expenditure of weapons.

Resource intensity provides a simple way to measure the capability of available resources to accomplish plan objectives. As with task intensity, resource intensity can be applied to an entire plan or used to reflect the relative level of resource allocation to an individual objective.

Resource intensity relates to the differences in resource capability needed to achieve a specific objective. In the example above, several different options are available for executing the selected actions. One could employ cruise missiles, stealth aircraft, or any combination of conventional aircraft. Each of these alternative resource assignments would require a different number of primary and support aircraft to accomplish the desired level of destruction.

There is also a time factor involved in determining intensity. The number and type of resources available for simultaneous operations limit the total resource profile for an operation. The key to effective planning and scheduling, then, is to match the tasks to the resource limits by reducing the intensity of an objective or by extending the time available to accomplish the objective.

A DIFFERENT APPROACH TO PLANNING AND SCHEDULING

Three features characterize the JPS approach. First, these technologies focus at a higher level of the overall planning and scheduling process than most of the other technologies discussed in this book. JPS employs artificial intelligence (AI) techniques, based on constraint reasoning and knowledge bases, to assist the decision maker decompose high-level objectives, such as gain and maintain air superiority, into specific tasks for action, such as eliminate a specific long-range radar site, or disrupt operations from a specific enemy airfield.

Second, JPS features a tight coupling between planning and scheduling. This approach to integrating planning and scheduling follows an objective-based paradigm instead of the usual resource-based paradigm. In other words, JPS seeks to maximize the number of objectives tasked (even if that means looking at alternative methods for accomplishment) rather than maximize the use of resources.

JPS is built with an understanding that there are many methods available to accomplish high-level objectives. Another benefit of tightly coupled plans and schedules is the elimination of the requirement to complete the entire plan before initiating scheduling activities. With JPS, planning takes place in an incremental process that divides the overall plan into segments, or sub-plans. Planner passes each completed sub-plan to Scheduler, which then assigns specific resources to the tasks contained in the sub-plan. In turn, Scheduler provides feedback to Planner concerning the aggregate availability of resources for subsequent sub-plans. If Scheduler cannot meet the constraints specified by Planner for specific tasks, then Planner and Scheduler enter a negotiation to modify the current and previous sub-plans in order to arrive at a mutually acceptable and executable solution.

The third feature of this technology is its dynamic ability to maintain and repair both the schedule and the plan. Both Scheduler and Planner maintain a common understanding concerning the constraints imposed on both the overall operation and the individual tasks. These constraints set the boundaries for both schedule and plan repair. Schedule repair, for example, occurs when an air strike is scheduled against a railroad bridge, but due to poor weather in the objective area, the strike is cancelled just prior to launch. Scheduler will then attempt to reschedule the air strike, as long as it falls within the time constraints set by Planner. Plan repair, for example, occurs when Planner, working with a human staff officer, determines that a plan objective will not be achieved by the designated time. Planner then determines that if the higher-level objectives supported will accommodate a delay, then it will recommend that a new time for completion be established for the original objective.

Repairs are performed within the constraints that originate with planning guidance. A staff officer, using JPS, gains a better understanding of the relationship between constraints and mission objectives. This understanding allows the staff officer to insert a padding factor in the plans and schedules to accommodate unforeseen events (like the rainstorm) that temporarily preclude the execution of specific tasks. Such a planning pad therefore allows Scheduler more flexibility to reschedule uncompleted tasks—as long as the revised schedule adheres to plan constraints. However, when Scheduler cannot satisfy plan constraints, it notifies Planner. Planner, in turn, determines if it can relax plan constraints or if a new or modified plan is necessary.

AGILE PLANNING AND SCHEDULING IN ACTION

Saturday, November 23, 2015, is a cold and blustery day in Washington, DC, previewing a long, cold winter in the nation's capital. There's a light dusting of snow blowing across the mall. Inside the White House Situation Room, and across the Potomac in the National Military Command Center, the activity level is high enough to make each room quite uncomfortable. Another in an almost-continuous series of crises has erupted. The Secretary of Defense is in contact with the Commander-in-Chief Pacific Command, who has already named the Commander, Joint Task Force (CJTF).

On board the USS Ronald Reagan, CJTF has assembled her staff to review the situation and develop alternative courses of action. The ship, currently en route to the conflict area, will commence operations in 36 hours.

Via satellite link, the planning staff receives the latest updates to the region's domain-specific files. This information includes the region's geography, the current order of battle, centers of gravity, and threats to both U.S. and allied interests. Accordingly, CJTF's staff updates JPS. Additional information regarding the current world-state comes from various intelligence sources. CJTF receives her initial guidance, then reviews high-level objectives with her staff. The planning staff begins to develop the operations plan. It's a team effort.

Human planners develop the plan through an interactive session with JPS' planning component. It's a perfect match, with humans providing overall guidance, while the technology quickly implements the various rules of engagement to present potential solutions. The staff officer, using a set of interactive and intuitive visualization tools to adjust individual objectives and constraints, fine-tunes the plan. This man-machine team is able to develop and test several high-quality alternatives in a matter of hours—even minutes—instead of days.

The choice of a COA is complicated by the fact that the U.S. is negotiating with several potential allies to undertake a coalition response—even while seeking a United Nations mandate authorizing the use of force to restore order. Coalition operations are important politically, but the addition of each coalition partner adds complexity in the form of new and modified objectives. In the late twentieth century, the planning staff would have been frantic, due to the constant changes in planning objectives. Now, the planning team members calmly and efficiently adjust the plan as they receive each change. They know that dynamic plan repair is just as important in the pre-execution phase as it is during execution.

Throughout the process, CJTF and her staff identify and modify objectives and tasks, responding to and anticipating changes in the evolving situation. The JPS tools support this interactive development of courses of action. What is exciting about this process is that JPS integrates both planning and scheduling to provide feasible, executable solutions in response to changes in the situation.

Approaching the area of operations, CJTF is informed that three ships from the Royal Australian Navy have been dispatched to assist. The staff on board the Reagan enters these new resources into the Joint Planner-Scheduler knowledge base. Scheduler retrieves the characteristics of these ships from the online database, then, based on this new allocation of forces, redistributes tasks to all units.

Upon arrival in the area, CJTF learns that the crisis has escalated. The aggressor nation has carried out several air strikes on merchant shipping that appeared to be heading toward disputed waters. The United Nations has formally condemned these actions, and CJTF, acting under the authority of the United Nations, is directed to eliminate the threat to unarmed merchant vessels. New objectives are added to the plan. Planner assesses the impact of these new objectives and notifies the staff that the plan must be repaired. Three of the seven original plan's sub-plans are recalled.

Scheduler provides updated resource availability, and Planner recommends four new sub-plans. The human planning team begins a systematic exploration of the alternatives presented to CJTF as part of a decision brief. CJTF offers several suggestions and adds new constraints, which are inputted directly into the Planner. CJTF and her staff observe the constraints' impact on the plan. CJTF then decides to avoid direct strikes against the airfields. With the addition of this constraint, Planner recommends that a surface-action group, consisting of two U.S. destroyers and three Australian frigates, be dispatched to the area of the attacks to escort neutral shipping. The plan calls for the Reagan to provide air cover for the naval escorts.

CJTF's staff officers are active participants throughout this process. They review recommendations, add new constraints, and assess the impact of the changes to the broader military and political context. For its part, the technology largely provides rapid and comprehensive recommendations. Staff needs are satisfied both by the detailed information contained in those recommendations and by the variety of alternatives JPS offers.

In the scenario above, human decision-makers interact with JPS, reviewing the predicted outcomes and schedules, and offering advice to modify execution. Throughout, JPS demonstrates a dynamic ability to adjust, repair and replan in response to new objectives. Partial plans, representing highest-priority tasks, pass from Planner to Scheduler for execution, and Scheduler re-assigns resources in response to these changes. Additionally, Scheduler recognizes the need for schedule repair. When, for example, the Reagan is tasked to provide air cover for the naval escorts, the crew determines that the ship should relocate to facilitate air operations in both directions. Scheduler originally schedules missions based on current location, but as the ship moves Scheduler will reschedule missions based on reduced time en route. Additionally, previously scheduled tanker missions may be eliminated, with the resources used to support other tasks.

With JPS, CJTF is quickly able to develop and present several fully developed and dynamically adjustable courses of action to the National Command Authority. What's more, when enemy actions or environmental factors result in delays or new tasks, she and her staff can keep pace with the dynamic execution environment. JPS' components work together with the human decision-makers to ensure that the COA remains current and appropriate.

JPS ARCHITECTURE

JPS' technical architecture provides for adaptable solutions to a variety of planning and scheduling situations. Planner, Scheduler, and Plan Server (see Figure 6-1) work together to overcome complexity and scalability issues facing military officers tasked with planning and executing the fast-paced operations demanded in today's world. Indeed, such tools extend and enhance the staff officer's capabilities. In this way, the human and the computer are a team, working together in an iterative manner to develop and execute quality plans.

JPS Planner

Planner performs domain-independent core planning based on Hierarchical Task Network (HTN) concepts for decomposing high-level objectives into specific tasks, including sequencing, priority, and mission-horizon information. A domain-specific knowledge base augments Planner's core planning capabilities [1].

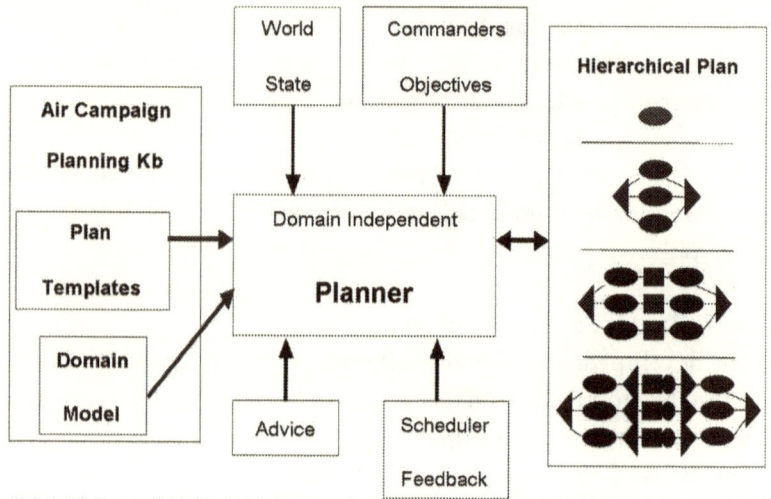

Figure 6-1. JPS Planner consists of a domain-independent artificial-intelligence engine employing a HTN methodology for plan generation and minimal perturbation repair.

Domain-independent Planner receives input from several sources, as shown in Figure 6-1. The Knowledge Base contains a set of process templates that guide the strategy-to-task decomposition process, and a model that provides specific information for the region and countries involved in the operation. Planner also receives updates on the current world-state, using an assessment tool to identify differences between actual and predicted world-states.

The planning process begins when the staff officer inputs the Commander's high-level objectives. Planner supports an incremental, interactive approach to planning. Working with a human counterpart, Planner develops the plan through a systematic decomposition of high-level objectives into a set of specific actions. Planner employs an AI planning engine, coupled with the domain-knowledge base, to select appropriate templates from the process template library. The process templates identify sets of actions that support higher-level objectives.

Working with Planner, a staff officer is free to select one or more templates from the recommended set. He can edit the templates either to eliminate specific actions or to add actions from another template. The decomposition process creates a complex network of objectives, actions, and tasks, all connected with many relationships, such as dependencies and time precedence. The process eventually reaches a level—the decomposition level—where the complexity of the tasks significantly diminishes. At the decomposition level, all subordinate tasks support one and only one higher-level objective. The decomposition level is the point where Planner defines the set of sub-plans that will be passed to Scheduler. Planner completes the sub-plans in order of their priority, highest-priority being the first. As each sub-plan is completed, Planner sends it to Scheduler for resource tasking.

Planner and Scheduler work in tandem to develop each sub-plan. Scheduler provides feedback to Planner as it completes each sub-plan, or if there are problems in resource assignment.

JPS Scheduler

Scheduler performs continuous, mixed-initiative allocation and scheduling of resources in tight integration with Planner. In addition to receiving the sub-plans from Planner, Scheduler receives periodic updates of the current world-state, particularly details of current and expected resource availability. An assessment tool—to provide a comparison of the actual state to the predicted state—processes these updates.

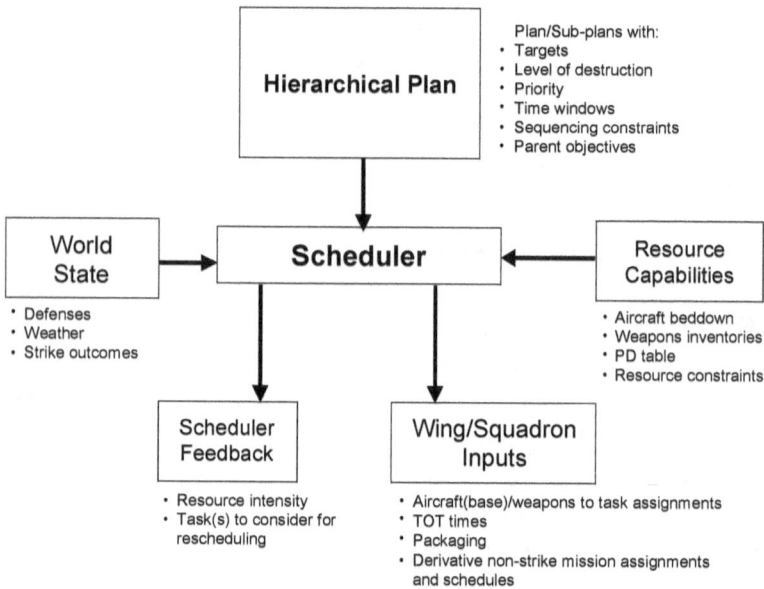

Figure 6-2. JPS Scheduler components provide a configurable system for reactive, mixed-initiative scheduling.

Scheduler produces specific tasking for the executing units. The tasking includes resources (amount and type), time for completion, packaging (grouping of resources for a specific mission), and support missions. Scheduler also provides feedback to Planner, recommending modifications to task intensity—that is, to reduce the intensity of selected tasks. For example, in the scenario, CJTF could eliminate the threat to merchant shipping by attacking the airfield supporting the enemy aircraft, or by establishing a defensive screen to protect the merchant ships.

Planner selects the airfield attack option, but due to limits in availability of attackers with the proper capabilities, Scheduler is unable to assign resources to accomplish all tasks for the particular sub-plan. Scheduler could then request a reduction in the resource intensity of attackers, which in turn would lead Planner to select an alternative approach—establishing the defensive screen. Increasing the allowed time for the task—if possible—might ease resource restrictions, which can be an effective method to eliminate a temporary problem.

Plan Server

Plan Server extends JPS' capability, allowing it to deal with large-scale operations. Through Plan Server, and using multiple, domain-specific knowledge bases, several specialized Planners can be linked to specialized Schedulers.

Plan Server helps manage the challenge of multiple, inter-related plans and schedules that typify military operations. Plan Server acts as a clearinghouse for all planning and scheduling actions. More important, Plan Server both maintains a set of triggers and monitors the system for the occurrence of trigger events, such as the failure of a specific task, or the violation of key plan assumptions. In the scenario, the attacks on merchant shipping represented an action running contrary to existing assumptions.

When it detects trigger events, Plan Server dispatches appropriate control messages to affected Planners and Schedulers, facilitating cooperative efforts among a number of specialized Planners and Schedulers for operations (air, ground, maritime), as well as logistics and communications in the iterative, interactive environment.

Deployment Configurations

The basic configuration consists of a Planner and a Scheduler. For larger operations, several Planners and Schedulers can connect through a Plan Server. Figure 6-3 shows potential configurations for small and large operational environments.

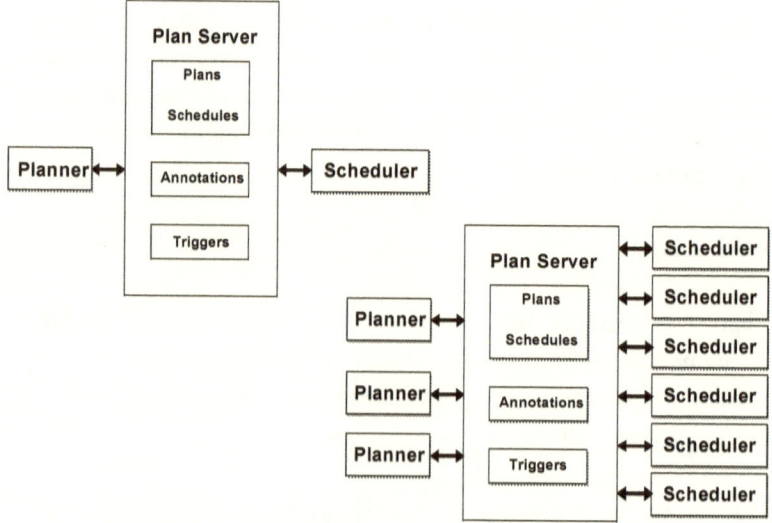

Figure 6-3. JPS can be configured with one Planner and one Scheduler for a simple operation, or with multiple Planners and Schedulers linked through a Plan Server for more complex operations.

COME TOGETHER: TRANSFORMING PLANS INTO ACTION

Perhaps JPS' most interesting aspect is its implementation of the very concept that motivated the nine officers described in the beginning of this chapter. They sought to improve C^2 by bringing together planning and execution. They knew that tightly coupled planning and scheduling begins with a sound conceptual framework. Then this conceptual framework must be implemented in the technology. Finally, the concept and technology require the support of a common representation of the objects and concepts that inhabit the domain.

Planner's interactive nature was presented earlier in this chapter. Here, the discussion focuses on the decomposition of high-level objectives into those specific, individual tasks passed to Scheduler.

Indeed, JPS' goal-driven approach differs from many other resource-driven planning approaches. By adjusting the intensity level needed to accomplish tasks, and using available resources, the goal-based approach seeks to accomplish all assigned tasks. While a resource-based technique seeks to optimize the use of resources, this approach seeks to maximize attainment of plan objectives [2]. The efficient use of resources is obviously still important, but the number of satisfied plan objectives defines efficiency.

Ready . . .

How does this technology maximize objective completion? Within a set of predefined limits, Planner and Scheduler interactively manage constraints until resources are assigned to all identified objectives. Planning, in the JPS context, begins with a set of high-level objectives: for example, support country A's efforts to repel an invasion and restore national sovereignty. Planner uses both domain—and scenario-specific knowledge bases to begin the process of decomposing this objective into executable tasks, such as conducting an amphibious landing, or gaining and maintaining air superiority. An example of a domain knowledge-base element would be a definition of tasks involved in escorting neutral shipping, while an example of a scenario knowledge-base element would be a list of a country's known radar sites.

Planning is an iterative and interactive process in which humans and machines cooperate to solve planning tasks. The human's role in this process is to guide the machine with advice. For example, Planner could suggest that air superiority be achieved over a certain large part of the objective area. The staff officer could counter by specifying a more limited area. JPS, then, occupies the role of a knowledgeable assistant to the staff officer, providing alternatives based on inputs made by the human. In the final analysis, however, the human is responsible for selecting between alternative courses of action, and fine-tuning the finished product.

Set . . .

This fine-tuning process takes several different forms. In the simplest context, fine-tuning may be nothing more than adding additional time between the desired completion of one sub-plan and the start of a subsequent sub-plan. Additional time could reflect expected periods of bad weather or a temporary shortage of critical munitions. Another example of staff officer fine-tuning is forcing a resource-intensive solution for a specific sub-plan at the expense of other sub-plans. If, for example, the enemy has a history of using submarines in operational exercises, in order to ensure that no submarines leave the port, from a strategic perspective it may be necessary to destroy all of the support facilities. Yet the same munitions used for this sub-plan might be required to attack C^2 bunkers. Here, the human member of the team is responsible for selecting the alternative to a specific sub-plan, while JPS provides a feasible solution for the rest of the plan.

High-level objectives typically map to a complex network of inter-related tasks (see Figure 6-4). A single objective may map to multiple tasks, and a single task will map to multiple objectives.

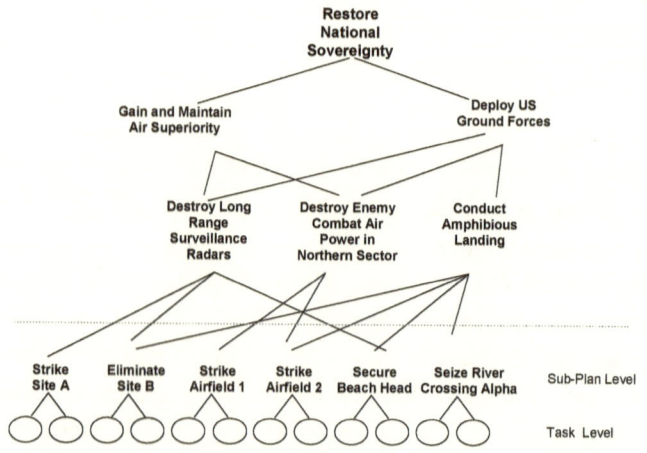

Figure 6-4. Objective decomposition iteratively breaks tasks down into lower-level tasks, eventually giving individual mission tasks that support a single objective.

Go!

This decomposition pattern typically continues to a point where all subordinate tasks map to one—and only one—higher-level objective. This critical point, called the decomposition level, represents the place where the plan splits into meaningful segments, or sub-plans. For example, in Figure 6-4, links from several higher-level objectives to the sub-plan call for establishing air superiority over the northern sector. This sub-plan supports the objectives to protect U.S. forces, establish U.S. ground presence, and disrupt enemy supply lines. Through the establish-U.S.-ground-presence objective, there is a relationship to conduct amphibious landing at location B, however, each of these sub-plans has its own independent tasks.

Planner both identifies a set of specific tasks to accomplish the sub-plan objective, and specifies a set of conditions or constraints that must be met while completing the assigned tasks. Here, constraints include time to complete, sequence (do this before that), and intensity level. This information is then handed to Scheduler for action.

MAKING INFORMED CHOICES: INTENSITY-BASED PLANNING AND SCHEDULING

The notion of intensity level is a key concept introduced earlier in this chapter. Here, the discussion focuses on the methodology used to calculate intensity level and how both Planner and Scheduler use this parameter to adjust individual tasks within and between sub-plans. Even though the individual tasks driving the resource requirements are not identified until after decomposition into sub-plans is complete, when defining individual sub-plans Planner requires some means of making informed choices between decomposition alternatives. Such a qualitative measurement—intensity—helps to differentiate between various options.

At the sub-plan level, Planner calculates on a scale of one to 10 the relative intensity of all possible options. For example, a sub-plan calling for the elimination of enemy C^2 communications capability could be accomplished by attacking various combinations of communications nodes. At one extreme, attacking a single, heavily defended node may disable the entire communications network. The same effect might also be achieved by attacking 15 more lightly defended nodes. Planner calculates the intensity of each of these options.

In the current prototype, intensity calculations are both specific to air operations and based on the number of targets introduced by tasks. In the example above, the single-node option introduces two communications targets and 10 air defense targets, while the 15-node option introduces 23 communications and 12 air-defense targets. Relatively speaking, then, the single-node option could receive an intensity ranking of two, while the 15-node option could receive an intensity ranking of six, denoting the fact that it introduced three times the number of targets.

This simple calculation would be adequate if there were a single class of resources available to conduct the missions associated with these tasks. Unfortunately, there are distinct differences between the capabilities of specific resources (aircraft and weapon combinations) to carry out specific missions. These resource differences are a critical factor in developing a feasible planning-scheduling solution; therefore, Planner also calculates the intensity level for specific resource categories for each option. Again, using the communications node example, Planner calculates the intensity for alternative resource groupings. Here, this calculation is not made for each aircraft and weapon type. Instead, based on values from the weapon effectiveness tables, the various individual aircraft types are grouped together. This approximation reduces the number of calculations required.

Using sortie-per-day tables, over time a maximum resource profile for the entire operation can also be calculated. Planner and Scheduler divide this resource profile by the number of sub-plans defined, thereby defining a desired resource profile for each sub-plan. By calculating the remaining resources available, and dividing by the number of remaining sub-plans, both Planner and Scheduler constantly track deviations from this desired level. Indeed, this constant monitoring is necessary to support direct intervention by the human staff officer. Again using the communications-node example, the human staff officer may decide that the 15-node option represents the best choice for political reasons, and therefore directs Planner and Scheduler to use that option. By reducing available resources, such a decision has an impact on all other sub-plans.

NEGOTIATING WIN-WIN

Many situations could lead to overtasking available resources. An aggressive plan, tight time constraints, and unexpected loss of resources can all lead to a situation in which the plan requires more resources than are currently available. When Scheduler is unable to comply with the associated constraints in assigning resources to a sub-plan task, it notifies Planner.

Scheduler assigns specific resources for each task of each sub-plan that it receives from Planner. In addition, Scheduler attempts to match the intensity levels specified by Planner, calculating deviations from desired intensity levels as it completes scheduling actions for each sub-plan. Scheduler then reports the specific resource intensity level assigned to the completed sub-plan, and recalculates the remaining intensity capability, to determine the new desired level for all remaining sub-plans.

Based on this updated information, Planner adjusts its selection of options. When resource constraints are identified, Planner can also adjust previously completed sub-plans. For example, in the amphibious landing force three Marine battalions could be assigned tasks to seize three specified river crossings. Because of a key communications node's proximity to the landing area, and a shortage of air resources, a subsequent sub-plan could call for a single Marine battalion to capture the node. However, timing constraints for capture do not permit replenishment of supplies between capturing the river crossing and seizing the communications node. Here, Planner could adjust the river-crossing task to reduce the number of crossings from three to two, then make the third Marine battalion available for use in seizing the communications node [3]. The sequence of these collaborative efforts between Planner and Scheduler is illustrated in Figure 6-5.

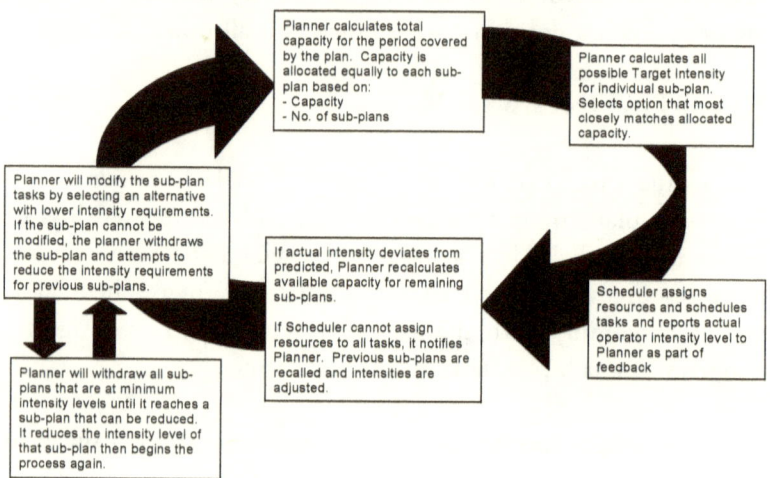

Figure 6-5. Planner and Scheduler work interactively, adjusting required intensity to match available capacity.

JPS' goal is to develop a solution that accomplishes all identified objectives. Thus, when Scheduler informs Planner that it is unable to schedule sufficient resources for Planner's requested task, the human staff officer, working with Planner, assesses the options available for constraint relaxation. On the one hand, he could extend the time for completion, either for the task in question or for other simultaneous tasks; on the other hand, he could accept Planner's recommendations to reduce the task's intensity.

In the river-crossing scenario, Planner, with the advice and consent of its human partner, could determine that one instead of three river crossings would acceptably support the objective of seizing the communications center. Here, other crossings could be delayed or eliminated, depending on overall plan objectives. By tracing the links among the objectives above the sub-plan level, Planner would assess the impact of these changes on all higher-level objectives.

For example, if a subsequent objective—calling for the introduction of U.S. ground forces into an internal area of the country—required a minimum of three river crossings, Planner, working in conjunction with the human staff officer, would assess various adjustments to overall plan goals and higher-level constraints. Tasks for the river crossings could be phased over time, allowing the use of the same units for all three operations. In addition, the follow-on operation, introducing troops into the center of the country, could possibly be delayed until all the crossings occur. As always, the goal is to discover a feasible solution that accomplishes all plan objectives.

MAINTAINING FLEXIBILITY

Planned actions always encounter unanticipated events that can disrupt the entire operation. Such disruptions often result in operations staffs discarding an entire plan and conducting the operation in an ad hoc manner until a revised plan can be developed. In general, military planning and execution staffs lack the ability to isolate an event's impact within the plan, in part because their planning systems are generally designed as monolithic processes requiring massive amounts of detailed information and considerable time to produce a detailed but fragile plan. Here, Joint Planner-Scheduler employs two concepts that help to overcome this deficiency: a dynamic repair capability and a dynamic open-ended process.

Dynamic Repair

The first concept, a dynamic repair capability for both the plan and the schedule, quickly adapts to unanticipated events with minimal perturbation. Based on changes in the real world-state or the Commander's objectives, the repair capability is designed to change only the plan's affected portions. These changes are accomplished by enabling Planner and Scheduler to guide the modifications made by each other—as a way both to reduce the time required to complete the adaptations, and to maintain greater stability (that is, fewer changes) in currently planned and scheduled activities. Indeed, the development of techniques that can support faster, less disruptive adaptations enables more agile yet stable C^2.

It is possible to consider modification approaches that either recompute the plan from scratch, taking into account the new information; or incrementally revise (or repair) portions of the plan, taking into account new information. For example, new Commander objectives may be quite disruptive, requiring recomputation of the plan; for example, if the Commander deletes nearly all existing objectives and adds a large number of new objectives. However, other inputs—for example, use weapon X against all type P targets—require changes only to those sub-plans that have type P targets.

Using either repair technique, an important issue is to find the effective level of information exchange between Planner and Scheduler. Here, transmitting all information is unnecessary, since it forces Planner and Scheduler to identify and discard irrelevant information. In particular, Planner should communicate to Scheduler only plan changes, and Scheduler should inform Planner of only those actions that are no longer supported in the modified schedule.

Open-Ended Process

The second important concept employed by JPS focuses on the plan as a dynamic, open-ended process. Here, JPS views a plan as a set of objectives decomposed into sub-plans and tasks; there is no beginning, middle, or end. This open-ended concept allows

JPS to adapt quickly to the incremental addition of new objectives. As such objectives are added, Planner begins the process of decomposition. By redetermining the desired intensity level across all current sub-plans, the system adapts to the addition and deletion of objectives at any level in the hierarchy. If the new sub-plans can be accomplished without changing resource allocations to existing sub-plans, then Planner will simply add these new objectives and sub-plans. When, due to constraints, new sub-plans cannot be accomplished, Scheduler will notify Planner. Planner will then systematically recall prior sub-plans and adjust the intensity level of each until a feasible solution is achieved.

This open-ended concept also supports incremental planning. Initially, only the first phase of an operation needs to be planned. As the operation unfolds, objectives and guidance for subsequent phases of the operation can be added. This method helps the Commander using JPS to conduct just-in-time planning, thus avoiding excessive replanning to reconcile a previous plan with unfolding events.

KICK-STARTING THE PROCESS

Many of the issues associated with implementing JPS technology are common to most C² systems. For example, data availability and format issues continue to be a perplexing problem. Multi-level security issues, involving the integration of intelligence and operational data elements, are being investigated in a number of projects.

However, there are also several unique JPS issues that need to be resolved before such an operational system can be implemented.

User Interface

In Planner, decomposition of high-level objectives and guidance follows a pre-defined set of decomposition rules. This system of rules and planning templates needs to expand to allow flexible human intervention in the decomposition process. A military operation typically unfolds in a series of phases, with each phase supporting one or more objectives. Such high-level objectives are further defined through a series of constraints based on geography, time, and rules of engagement.

The challenge, therefore, is to design an interface that supports visualizing the interactions among plan constraints, sub-plans, and intensity levels. A military staff officer must be able to see the impact of incremental changes to each of these factors on the overall plan. How is the set of sub-plans displayed? What do constraints look like? How does the addition of a new objective affect the allocated intensity across all sub-plans? Can a staff officer adjust the constraints to minimize the impact? Does JPS provide the solution? Clearly, the interface between a staff officer and JPS needs to provide visualization mechanisms that support finding and understanding the answers to all of these questions.

Data Availability

JPS requires finished intelligence data—that is, inputs from all available sensors and reporting agencies have been collected and fused into a single input representing the current perceived truth for the battlespace. The data issue is not unique to the JPS; current C^2 systems have similar problems.

Efficient use of JPS technology will also require mechanisms to facilitate automated world-state data updating. Specific tasks will have attributes, such as identification, location, status and value. As uncertainty occurs, there needs to be a factor that reflects the reliability of the available intelligence information—and this factor will require an infrastructure that provides a consistent vision of reliability. For example, following a strike on a specific target, feedback—in the form of pilot reports—may provide the initial indication of strike results. However, the validity of these reports may be reduced by limited visibility in the objective area, the use of decoys, or other adversary deception methods. If JPS reacted too quickly to feedback, such actions could cause JPS to implement unnecessary changes to both the plan and schedule. Additional information could then cause the system to issue new changes. Therefore, the result of several iterations could find the system thrashing to respond to each new information update.

There are a number of interfaces required to support the injection of operational and resource information into JPS. Data concerning resource and asset status, and availability, is currently reported through maintenance and possibly operations channels. Generally, data is collected as it becomes available at the local (facility) level, but only reported up the chain of command on a periodic schedule. To support JPS requirements, so that the system can make asynchronous updates on an as-required basis, there may need to be adjustments to the reporting period, or distributed access to resource data would have to be provided at the source.

Supporting the Life Cycle

There are significant life-cycle support issues involved in the development and maintenance of the plan/process domain templates. These templates represent the decomposition of high-level objectives into multiple levels of tasks contributing to the achievement of overall objectives. Effective use of Planner calls for a capability to develop and maintain such domain templates. Template creation and maintenance should ideally incorporate a description of the relationships among the various elements of the templates. For example, the relative priority of sub-tasks must be determined and documented. In addition, templates should be able to recognize complimentary objectives/tasks and mutually exclusive objectives/tasks.

The development and maintenance of domain templates is a long-cycle activity that would probably require intensive efforts for initial generation, but would only need updates every few years. During military operations, each service would provide a dedicated team of analysts to examine the results from the field, assessing the effectiveness of employed military doctrine and strategy. Interim updates and modifications to the doctrine templates could be issued, and then immediately sent via satellite to the appropriate commander. In order to measure success in achieving the objective, domain-template creation and maintenance should also define what information should be collected.

The capability to examine specific template compositions, and then to modify those compositions in response to theater constraints, is also required. To support the concept of developing an operation as a series of phases, each with its own set of objective templates, mechanisms for developing and storing template compositions for future application would be required as well.

SUMMARY

Systems like Joint Planner-Scheduler (JPS) attempt to overcome automation challenges in planning and scheduling in a manner that supports execution in dynamic environments. Planning, here, is a process of determining tasks necessary to achieve given goals. Scheduling is a process of assigning specific times, assets, and resources to the tasks—subject to task and resource constraints. In typical computerized solutions, the two processes are distinct in nature, and are difficult to integrate—a difficulty compounded when dynamic modifications of plans and schedules must be performed in the face of change.

Using the AI planning technique—Hierarchical Task Network— JPS' Planner decomposes higher-level goals into tasks, tasks into sub-tasks, and so on. (A domain-specific Knowledge Base is required.) Sub-plans are then passed on to Scheduler, which assigns specific times, assets, and resources to those tasks specified by Planner. As tasks are executed, Assessment tools compare actual situations to those predicted and inform Planner and Scheduler.

In some cases, Planner may establish a plan for which Scheduler cannot find a feasible schedule. To enable the two to communicate in a mutually comprehensible manner, JPS introduces a semi-qualitative concept—intensity—as a measure of resource usage. Planner looks for task decomposition options, with intensity consistent with available resources. Scheduler communicates the unfeasibility of certain plans to Planner, also in terms of intensity, so that Planner can replan accordingly.

JPS dynamic repair is based on the concept of minimal perturbation—reducing the amount of unnecessary disruption to execution, balancing agility, and stability. When unforeseen events occur, Scheduler attempts to repair the schedule within the existing plan and constraints. If such a repair is impossible, Scheduler notifies Planner, which then looks for a different plan—within the intensity constraints advised by Scheduler.

CHAPTER 7

KEEPING OPTIONS OPEN

'Joshua and the whole army moved out to attack Ai. He chose thirty thousand of his best fighting men and sent them out at night with these orders: "Listen carefully. You are to set an ambush behind the city. . . . I and all those with me will advance on the city, and when the men come out against us . . . we will flee from them. They will pursue us until we have lured them away from the city. . . . So when we flee from them, you are to rise up from the ambush and take the city." The Bible, Joshua 8.3-9'

The plan worked. Joshua's contingent attacked and then let themselves be driven back. The men of Ai were lured away from the city to pursue the Israelite army, confident of certain victory. But then the men in ambush walked into Ai and set the city on fire. The trap had been set and sprung.

'The men of Ai looked back and saw the smoke of the city rising against the sky, but they had no chance to escape in any direction, for [Joshua's army] turned around and attacked the men of Ai, [and] the men of the ambush also came out of the city against them, so they were caught in the middle, with Israelites on both sides. Ibid, 8.12-23'

The army of Ai made a classic, tragic mistake: they put all their proverbial eggs into one basket. By the time they realized that they had over-committed to a severely incorrect assessment of the enemy's position, their options were gone.

The fatal story of Ai raises a timeless question: How can commanders keep their options open, while still exploiting apparent opportunities?

EVALUATING THE FUTURE

Using a predictive model that anticipates important possible future contingencies, the effective commander can evaluate alternative actions. Clearly, an inadequate predictive model that fails to include some real contingencies can lead to disaster—it's possible that the leader of the Ai army was simply naïve, never entertaining the possibility of an ambush.

But being aware of all relevant battlespace possibilities is just the first step in successful C^2. Perhaps the Ai commander did consider the possibility of an ambush, but decided it was very unlikely. After all, Ai had already defeated Joshua's army once before. Nevertheless, ignoring the possibility of an ambush, however unlikely, was the commander's last decision. He realized too late that simply responding to the most likely scenario isn't necessarily a winning strategy. If improbable events will lead to defeat, they must be taken seriously!

USING FEEDBACK: THE VALUE OF FUTURE INFORMATION

Even if one scenario is clearly the most likely to occur, the savvy commander carefully evaluates how to respond to each of the possible futures in a predictive model. Future actions will depend on which of the future events actually occur. In control terminology, the commander develops a feedback strategy; that is, a strategy for mapping future information into future alternative actions. The commander wants to select the next action in a way that takes advantage of the most likely scenario while retaining flexibility to respond with success to alternative future events.

Future information becomes the key to success. Indeed, some actions in a feedback strategy are taken for the sole purpose of obtaining new information. A commander then can defer resource commitments until adequate information is available for him to make the best-informed decision.

A serious problem facing today's commanders is paradoxically too much information. A commander has access to an enormous amount of detailed feedback information from modern warfare systems. Yet no matter how good or effective the commander, humans can review only a limited number of likely scenarios. Consequently, today's commanders use high-level predictive models that focus on aggregate movements of friendly and enemy assets to deal with the complexity of modern battles. This chapter describes an alternative approach, a new technology that leverages the power of the computer to develop feedback strategies based on detailed predictive models.

COMPUTING FEEDBACK STRATEGIES

It's easy to envision using a computer to evaluate feedback strategies based on many more scenarios than a human commander could possibly consider. Moreover, the predictive model used by a computer could directly employ detailed battlefield data.

This is precisely what is done in the *Dynamic Aerospace Execution Tool* (DAET—pronounced day'-et), an application of a technology being developed by ALPHATECH, Inc., Burlington, Massachusetts, and collaborators from Boston University, Boston, Massachusetts. DAET incorporates real-time intelligence data into detailed predictive models to generate battle plans, and, using simulation, evaluates feedback strategies. Using probability distributions for individual events in the model, DAET also generates scenarios, averaging performance over hundreds of alternative future events.

DAET makes battle management more flexible and responsive. Given an update in battle status¾information regarding all assets and known/unknown threats and targets¾DAET evaluates and optimizes the plan for a 24-hour horizon in less than fifteen minutes.

FEEDBACK STRATEGIES VERSUS ROLLING HORIZON PLANNING

Generating a true feedback strategy, DAET explicitly takes into account the effects of future feedback information. DAET stands in contrast to the standard rolling horizon approach that uses current feedback information to update plans, but fails to represent the tree of responses that will be taken when future information becomes available.

To understand more precisely the difference between a DAET feedback strategy and the standard rolling horizon approach, let us consider a game of chess. With the two sides Blue and Red, let their respective k^{th} moves be denoted by b_k and r_k. Suppose further that Blue's k^{th} move is to be selected by looking ahead three moves; that is, the planning period for selecting b_k will take into account the next response of the opponent r_k, followed by the next move for Blue b_{k+1}, followed by Red's response r_{k+1}. In a DAET approach, this look-ahead procedure takes into account explicitly how Blue would respond at step k+1 to the move r_k taken by Red. That is, DAET evaluates a tree of move sequences, with b_{k+1} depending on the value of r_k.

To evaluate such a tree stands in contrast to the so-called open-loop feedback strategy that would result from the standard rolling horizon approach. In the latter case, a plan consists of a specification of b_k and b_{k+1}—without allowing for different values of b_{k+1} based on different values of r_k. At time k, the rolling horizon plan computes a single b_{k+1} as the best response to some expectation of r_k, that is, the expectation of what the opponent will do in response to b_k. Of course, the rolling horizon plan will be reevaluated following move r_k, so the planned b_{k+1} can be changed. Nevertheless, it is clear that there are situations where the move b_k at time k would be different—if the evaluation of that move included the fact that the next move, b_{k+1}, will indeed differ depending on the value of r_k.

Since DAET strategy includes the move sequences that are considered by the standard rolling horizon strategy (in other words, DAET includes an evaluation of plans for which b_{k+1} is the same for all possible r_k), DAET strategy will always do at least as well as the standard rolling-horizon strategy.

DAET IN COMMAND AND CONTROL

If DAET had been available to the Ai army commander in Joshua's day, it might have played out the ambush scenario in its simulator and recommended holding adequate forces in reserve to defend the city. DAET may have even recommended a reconnaissance mission to determine quickly whether or not an ambush was present.

The Ai scenario is admittedly simple. Let's consider the impact DAET technology could have on highly complex modern C^2.

The Situation

Senior officials decide that a preemptive attack is necessary to respond to threats from unfriendly forces, and they will use DAET to generate mission orders for the Air Forces Commander (AFC). As the situation unfolds, DAET uses two key capabilities:

— Dynamic Generation/Modification of the mission orders. Reflecting the arrival of intelligence data, DAET generates and continuously updates the mission orders. This process ensures that mission plans are responsive and feasible for current battlespace conditions.

— Proactive Assignments. At first glance, dynamically generated mission orders may appear to have suboptimal mission assignments; however, further analysis indicates that DAET strategically reserves assets for potential future contingencies that might have been otherwise overlooked.

While the first capability is a feature of standard rolling-horizon technology, the second demonstrates the full strength of computing feedback strategies that take into account a wide range of possible responses to changing circumstances.

Let's see how the battle unfolds.

Pre-Battle Plan

Senior officials develop the strategy and overall objectives. Based on the strategy, the staff decomposes enemy territory into logical sectors divided by geographical, strategic, and tactical considerations. Through database commands, operations (Ops) planners instantiate a DAET model for each logical sector and load data for each instantiation. In part, the data includes staff-developed target value assignments, some of which increase or decrease with time, reflecting the relative values of destroying targets over the planning period.

Initialization

An Operations (Ops) planner (a fighter crewmember trained to use DAET) takes responsibility for each sector, using a custom graphical computer interface to manage the DAET system's data and mission generation. The Ops planner:

— ensures that the relevant data is loaded, including the latest parametric and situational (geography) data

— connects to the intelligence division in order to obtain information as the battle unfolds, including updates to the Enemy Order of Battle (EOB)
— connects to bases allocated to their chain of command to determine the Friendly Order of Battle (FOB) available (Note: Some bases may not be in the theatre.)
— connects to the logistics center back in the U.S. to determine inbound assets to improve or sustain FOB

Let's follow the activities of one Ops planner managing the northern sector.

Day 1—Initial Tasking

Senior officials hand down the initial determination of targets and their decomposition, which are then loaded into the DAET scenario generator. In this scenario, the first day will consist of three waves of aircraft attacking en masse. The Ops planner uses DAET to generate the initial flight packages. The DAET simulations indicate the following:

— Current intelligence data indicates that only surface batteries that include guns and missiles protect the northern targets. In such an environment, with limited enemy airborne defenders, strike aircraft will be able to defend themselves. Therefore, there is no need for support fighters to escort packages.
— Remaining airborne defenders can be diverted to packages in the other quadrants and friendly airfield protection.

Evaluations

The AFC's upper level staff reviews the initial set of mission orders to ensure that it reflects the strategy that they have defined. They approve the initial tasking and authorize its release to the wings. The Ops planner sends the tasking down.

Wing mission planning cells review the tasking against their assets and available munitions. The wing personnel note that the munitions assigned to a target are not as effective as some other munitions available. They query the AFC as to the reasoning. AFC, in consultation with the Ops planner, notes that a higher-priority target will need those limited munitions in the third wave or the percentage of kill won't be achieved. Therefore, the weapons are being saved until then. Additional specialized munitions are not scheduled to arrive until day three, well beyond the time needed to hit the target. Thus the tasking stands. Therefore, DAET correctly supports future requirements.

The wing is also concerned that the Weasel support may not be sufficient. The wing looks at its resources and determines that another Weasel is available and can be used to decrease enemy defense density in the target area. The AFC approves the change. This was not overlooked by DAET, because the availability of the additional Weasel had not been included in DAET data.

The wing also notes that a flight of Weasels is to be put on ground alert in case a pop-up SAM (surface-to-air missile) shows itself and threatens an approach corridor. It concurs with this precaution since intelligence is limited during the initial phase of this war. Therefore, DAET correctly anticipated a future contingency.

Wave 1

The wing downloads the inputs from DAET flight packages into its local mission planning system, which gives the mission commanders the data needed to plan the tactical portion of the mission.

During the preparation and execution of the first wave, new inputs arrive from intelligence and logistics, updating EOB and FOB. The Ops planner runs DAET with this new data and determines that it is not necessary to reassign any missions or initiate new flight packages.

The Ops planner has time to vary data and determine which parameter variations will have a large impact on the ongoing plan. Employing forecasts of areas of potentially significant events, the Ops planner uses DAET to vary the data and reevaluate the plan. By recording which variations do not cause changes in the plan, and which lead to significant changes in missions and flight packages, the Ops planner is better prepared to respond to questions from AFC and wings concerning possible DAET-recommended reassignments and changes in the plans. Therefore, DAET facilitates what-if studies.

The Ops planner is also able to view graphically how individual DAET simulations evolve, gaining insight into why the simulations reacted as they did. Any major revelations will be brought up with the AFC.

Wave 2

When feedback is received from the first wave, the Ops planner runs DAET to plan the second set of flight packages. The previously cited plan generation, evaluation, and execution cycle is again carried out.

During the second wave of the attack, intelligence identifies the possible position of a targeting radar, associated with a surface-to-air mobile missile system not previously known to exist. The Ops planner notes this on his display and brings it up to his superior via the main viewing board. His superior, the Chief of Operations, looks at the location of the possible radar site, and the effect this site might have on the progress of the campaign, visualizing both by playing DAET simulations on the big screen. With DAET including the possible threat as real, he determines that this radar site could be a deterrent to the campaign and orders further investigation.

While the Chief of Operations and the Operations Planner were evaluating the situation, DAET, running simulations in the background, develops a recommendation to task a UAV to this location. The Ops planner notes a UAV is orbiting close by and selects it with his mouse. Data from the intelligence database appears in a scroll-down insert window, including the current state of the UAV and its capabilities. The phone rings; a military intelligence specialist is curious as to why the Ops planner is looking at the UAV. Discussing the situation with the military intelligence specialist, the Ops planner determines that there is no intelligence on the site and that this UAV is perfect for the mission. Therefore, DAET, based on the most recent intelligence data, took into account the new threat, and the new resource, and replanned the mission. Accordingly, the Chief immediately assigns the UAV to the new mission.

During the UAV's flight to the reported threat, the Ops planner receives post-strike data, updated intelligence, and updated logistics flow. Using this new data, he runs the DAET simulations, making note of recommended changes to the last attack wave. He categorizes these recommendations and e-mails them to the Chief, who will bring them up at the operations meeting, scheduled to begin shortly.

The Ops planner notes that the UAV is communicating its status and that this information is being data-linked to military intelligence. A short time later, a synthesized flow of information appears on his screen. Additionally, a recommendation from the intelligence specialist also appears, saying that there is a 90 percent probability that there is an active SAM system in the specified area. DAET has run ingress and egress simulations, making sure it is possible to avoid other strike packages and known threats. Via data-link, the Ops planner notifies ground alert that their Weasels now have a target. In addition, the Weasels are also given location and frequencies for a post-strike tanker, if needed. Finally, they are given the position of the on-site UAV that will be monitoring the attack and providing post-strike data to the command post and the wing. So alerted, the Weasels launch.

Wave 3

While the Weasels are proceeding to the target, the AFC receives his daily briefing, beginning with an intelligence update, post-strike evaluations, and discussion of the next strike. Halfway through the meeting, the AFC notes that the alerted Weasels are arriving at their target location. He directs the Ops planner to call up the UAV live video¾which he does just in time to see their HARM missiles destroy two mobile SAMs.

Pleased with the results, the AFC continues the meeting. Based on the attack waves' results, they discuss possible changes to the plan. Following DAET's recommendation, the AFC suggests a transition to continuous pressure attacks, revolving around the weapon systems' available mission cycles.

Fighters will be turned the most, while tankers and bombers will take the longest because of the required time en route. Special forces will be deployed via submarine to destroy coastal radar, thereby reducing the accuracy of anti-ship missiles being launched against the carrier battle group sitting off the northern coast. Facing a much-decreased risk, the battle group will be able to move closer to the coastline safely, reducing the need for tactical tanker aircraft for the F-18s. Therefore, DAET correctly took into account the times needed for all these maneuvers.

The AFC tells his staff to continue to monitor the situation and advise him of any major deviations to the plan.

A CLOSER LOOK

This battle story illustrates DAET's principal capabilities, demonstrating how the technology can anticipate likely contingencies and position assets for unexpected opportunities. A key point is that DAET's anticipatory behavior, consistent with current air operations, is not a trait of standard rolling-horizon planning technologies. Instead, in many ways DAET emulates the capabilities of the best military strategists, but it does so at a frequency and volume that only a computer can achieve.

Figure 7-1 illustrates a DAET application. A trigger event—aircraft attrition, enemy assets destroyed, change in strategy/goals, new information on threats or battle targets, change in friendly asset availability, mission completion—initiates DAET execution. Using updated information from the battlespace, along with data characterizing assets, threats, and targets, DAET generates new missions for the air assets. Based on a 24-hour planning horizon, these new missions are delivered within 15 minutes of each triggering event.

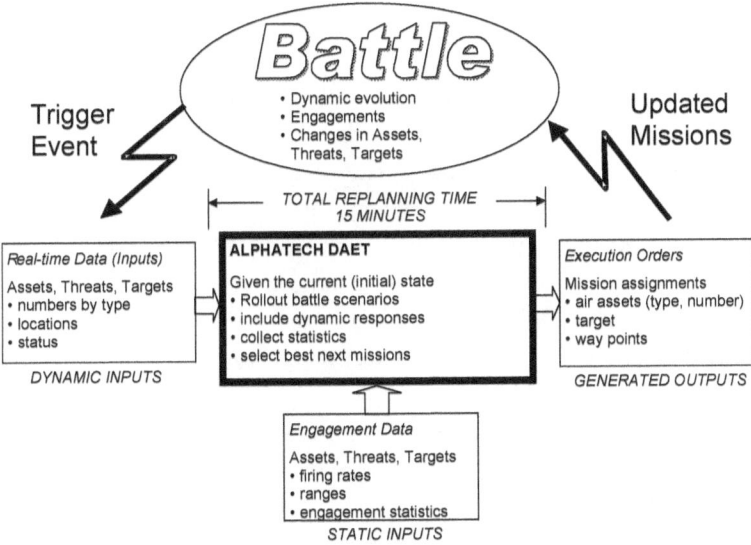

Figure 7-1. The DAET planning cycle takes into account real-time and fixed data inputs when evaluating alternatives and generating orders.

A DAET-generated air mission consists of flight package composition, targets, waypoints, routes, and weapons. Over the 24-hour planning horizon, and using several sources of information, including available assets (number, type, values, and so on), known threats and targets (estimated location, values, and so on), and intelligence data on emerging targets and threats, DAET creates sets of missions. Probability distributions for possible individual events help create simulated evolutions of friendly and enemy assets. Simulating the battle—starting from the current state and evolving over the 24-hour planning horizon—DAET evaluates alternative missions. Here, the discrete-event simulation engine executes the battle dynamics predictive model, rolling out the battle scenario multiple times, with each simulation scenario corresponding to one realization of all of the predictive model's probabilistic events. The results are averaged over the multiple simulations to optimize the mission plans.

Predictive Modeling

To manage air assets, DAET models the battlespace with the following four types of objects: airbases, flight packages, threats, and targets. Friendly assets not currently involved in missions reside at the airbases, which in turn specify the number and type of each available aircraft. Flight packages, comprising aircraft, targets, and routes (a sequence of waypoints to targets and back to airbases) define missions. Each flight package moves through the airspace according to the dynamic characteristics of the particular aircraft.

Threats and targets characterize enemy assets. Engagements with the enemy occur when the flight packages are within the ranges specified for threats and targets. Probabilistic models determine the outcomes of engagements (aircraft losses, target and threat kills).

The dynamic evolution of friendly and enemy assets is characterized by sets of variables defining the positions of flight packages and their physical constraints, such as aircraft velocities and range (where range varies with time as fuel is consumed). Aircraft in flight packages and enemy assets (again, threats and targets) are tagged as alive or dead. These variables change per the probabilistic models of engagements, and, in the case of targets and threats that appear only during specified windows, per time. Available weapons and firepower are also specified by variables that change as missions are carried out in the battlespace.

Given the current states of all friendly and enemy assets in the model, the evolution of the battle is simulated forward in time, integrating aircraft dynamics to update positions, and computing the results of such events as engagements. This simulation leads to a model that can be formally characterized as a Markov decision process, where the decisions (controls) at each time instant are the compositions of the flight packages (aircraft, waypoints, targets). Similarly, the available controls at any instant are constrained by the current state of the assets, specifically, the assets and missions in progress.

Algorithms

Using current state information, DAET initializes each simulation. Several schemes are used to partition the optimization problem, reducing the size of the decision space for selecting the missions [1].

DAET's principal distinctive feature is that it implements planning strategies that better account for the way battles actually occur— plans change as new information becomes available, and these changes take part in decisions about what should be done next. The optimal algorithm for solving sequential decision problems of this type is known as dynamic programming (DP). Yet because of the explosion of possible states and responses, it is impossible to implement such a complete dynamic programming algorithm. Consequently, DAET uses recently developed techniques to reduce the search space, thereby obtaining good approximations of the optimal solution [2].

An approximate DP strategy will always do at least as well as a standard rolling horizon strategy; however, certain situations may not warrant the added computational cost of even an approximate DP solution. In such cases, DAET will switch to standard rolling-horizon strategies, with varying degrees of sophistication and computational complexity, which can be tailored to the current battlespace situation.

One significant issue in solving optimal sequential decision problems concerns the way the environment is modeled. In the case of battles (and the scenario described above), the environment includes an adversary who is also making decisions and undertaking actions. If the adversary's decision process is represented explicitly, the problem becomes a game. When that occurs, game theory must be invoked—and then, while simply defining what is meant by a solution can be difficult, computing a solution may be impossible. Therefore, to avoid the complexities inherent in game formulations, DAET instead models opponents in the probability distributions of engagements. Such a model captures well the enemy's defensive actions—if flight packages come in range of enemy threats, for example, an engagement ensues, and losses are accrued according to the players involved and the static inputs characterizing the results of such engagements.

Inputs and Outputs

As shown in Figure 7-1, there are two types of inputs to DAET. Static inputs are data characterizing the features, capabilities, and values of assets, threats, and targets. Such inputs also include probabilistic characterizations of engagements between various pairs of friendly and enemy assets. Dynamic inputs are data that is updated as the battle progresses, including the locations and status of friendly and enemy assets. DAET uses both sets of inputs to prioritize and select missions, initialize and execute the simulations, and evaluate results—all to generate updated mission recommendations.

DAET produces missions, or flight packages, comprising assets, target objectives, waypoints (route to and from the target), and weapons. As a battle evolves, missions are executed by traveling to the waypoints in sequence, engaging the target when reached, then returning to the base along the route designated by remaining waypoints. When the flight package is within the range of a threat, the resulting engagement can mean the loss of friendly as well as enemy assets.

In every case, DAET selects waypoints according to two things: the geographic distribution of threats and target locations, and a simple shortest-path algorithm, where shortest is defined in terms of threat costs along the path. DAET also evaluates mission alternatives according to results from multiple simulations, and not merely from an evaluation of a particular waypoint selection's cost.

When DAET makes an evaluation, missions underway may be reassigned, either retasking to a new time-critical target, aborting and returning the flight package to its base, or retasking surveillance aircraft to a new target area. Throughout, DAET can also factor in costs for reassignments to avoid inappropriate thrashing, that is, unproductive repeated mission assignment changes.

DAET generates a set of immediate missions, not a schedule of missions for the complete 24-hour planning horizon. Indeed, DAET does not compute such a day-long schedule since the uncertainty of the battlespace would likely render any particular set of future missions infeasible or inappropriate. Actually, the idea of computing such advance missions contradicts the basic DAET algorithm philosophy, which is to evaluate feedback strategies rather than fixed mission schedules (as in conventional rolling-horizon planning). Nevertheless, DAET is not prohibited from producing a complete strategy. If DAET were to generate a 24-hour strategy, it would deliver a set of sequential responses to all possible triggering events that might occur in the battle.

THE QUESTION REMAINS

Using computers to construct battle plans sounds appealing because, employing the enormous amounts of data available in today's battle management systems, computers can tirelessly evaluate many alternatives. Nevertheless, before such technologies as DAET can be embraced there are serious issues to be considered, all relating to the data required to populate the predictive models used in the evaluation of feedback strategies.

Existing Data

Much of the data required by a technology such as DAET is currently available, including the EOB, FOB, and parametric data characterizing friendly and enemy assets. But this data needs to be made available in a way that can be used effectively in predictive models. Toward this end, the following issues need to be addressed:

— Developing a robust infrastructure that will allow the seamless exchange of data between multiple heterogeneous data sources and the predictive model.

— Translating the data into a form required by the predictive model. This requirement can be met through a variety of emerging and maturing technologies, for example XML. However, the pitfalls of data translations (for example, consistent units of measure) are well known. One of the major challenges is that the same information may reside in more than one data source in more than one format.

— Recognizing changes in the dynamic data and correctly updating the predictive model. Mechanisms for recognizing change, in the form of triggers, are part of most commercial databases. While it is likely that human expertise will be required to assure the updates are merited and correct in many cases, the challenge will be to filter events so that necessary human intervention will not become a bottleneck.

New Data

Capturing the desired objectives for a battle, so that feedback strategies can be optimized, is the biggest challenge to the adoption of any computer-based battle planning. Currently, the process for target valuation is manually intensive and highly dependent upon the Ops planner's subjective understanding of his commander's guidance. Given the high level of subjectivity in the current process of target valuation and prioritization in a medium-to-large conflict, it will be difficult for a computer to arrive at an optimal plan for achieving a commander's objectives.

Therefore, more research is required into other mechanisms for capturing the commander's intent (in other words, mechanisms other than target valuations). Given more elaborate modeling, a predictive model would be able to generate optimal strategies with respect to any objective. For example, the objective could be stated in terms of a desired end state (such as, at least some percentage of targets destroyed), and with this objective the computer would generate an optimal strategy by measuring the estimated end state relative to the desired end state.

The important implication is that there must be a means for capturing and representing the commander's guidance and objectives. In this regard, DAET is more flexible than many other proposed technologies because its simulation-based analysis does not require that the objectives be mapped into one particular representation (such as target valuations). A better way of representing the desired outcome needs to be developed, and a better, quantifiable measure needs to be identified. It is also certain that new data sources will have to be created to capture such information, since there are no current technologies that require it.

In summary, although computer-based generation of feedback strategies is a promising technology, the infrastructure and data required for the predictive models are not yet available. Research projects are currently underway, however, to develop the necessary information infrastructure. Thus, the prospects for deploying DAET-type technology in the future are good.

The Option Not Considered

When resources are limited, and the future is uncertain, a commander needs a strategy that retains sufficient flexibility to respond to future contingencies. In control terminology, the commander needs a feedback strategy. Other approaches to planning cannot fully exploit the value of future information.

Emerging technology makes it possible to evaluate alternative responses in full detail to an ever-increasing number of possible futures. Using the latest detailed intelligence data, the computer can assess many scenarios, in contrast to the human commander, who must rely on subjective assessments and quick judgments to reduce the search space to only a few alternatives.

Computer-generated feedback strategies can anticipate contingencies at a frequency and volume that will never be possible for human commanders, and strategies developed by prototype technology have proven to be operationally consistent. Moreover, computer-based predictive models make it possible for human commanders to play what-if games based on the latest parametric and battlespace information. The risk of not adopting such technology is clear. If the enemy has options that you haven't considered¾as Joshua did with Ai—they may be the ones that lead to your defeat.

SUMMARY

With limited resources and an uncertain future, a commander needs a strategy that retains sufficient flexibility to respond to unforeseen contingencies. In control terminology, the commander needs a feedback strategy. Merely planning for a most likely scenario (or even multiple ones) can lead to defeat.

A feedback strategy maps future information into future alternative actions, allowing a commander to select the next action in a way that takes advantage of the most likely predicted scenario while retaining sufficient flexibility to act. Using many more scenarios and more detailed data than a human could manage, a tool like DAET can compute feedback strategies. Going beyond a standard rolling horizon approach, DAET explicitly takes into account future feedback and considers a tree of responses. DAET's basic philosophy is not to compute a fixed schedule of tasks and missions, but instead to compute feedback strategies for agile actions. For example, the scenario in this chapter describes how DAET can anticipate likely contingencies and reserve appropriate assets to meet those contingencies.

A trigger event initiates DAET execution. Using a predictive model of battle dynamics, the algorithm probabilistically rolls out multiple battle scenarios, averaging the results over multiple simulations to optimize battle plans. A probabilistic model is used to determine the occurrence and outcomes of engagements. The simulation is a Markov decision process, in which decisions at each time are compositions of flight packages.

To avoid computational intractability, instead of using a complete dynamic programming algorithm, DAET uses a less computationally expensive approximate technique. To avoid complexities of a full game formulation, in which an adversary's decision process is modeled explicitly, DAET models the adversary through probability distributions of adversarial actions.

PART IV
OPERATIONAL COMMAND AND CONTROL

CHAPTER 8

QUANTITATIVE MANAGEMENT OF UNCERTAINTY

It's 0600 hours, and you've just received your tasking. You have until 1800 hours tomorrow to clear a path for a major offensive starting at 2100 hours. Your target array consists of multiple radar sites, missile batteries, and a couple of hardened C^2 nodes.

Higher authority sees your task as very important. Your task, therefore, states a required probability of success of .95, a conventional part of task definitions since the year 2007. This current state of task definitions means that higher authority will accept, in the long run, a risk of failure of only five times out of 100. Higher authority is also sensitive to casualties. Because the assets currently under your control may be needed for future operations, if you lose more than three platforms you might as well consider your task a failure.

For some time now, you've been planning to use your remaining precision-guided munitions in your two other tasks, already underway. Now you think, perhaps you should use them for the new task, since it must be finished prior to the other two. If you do use them, will you execute your new task successfully? Since you can never know with complete certainty, you must decide whether using these munitions is a reasonable move; more specifically, you ask whether the risk associated with this alternative is acceptable.

If you do use your more sophisticated weaponry for your new task, would you be exposing your other tasks to unacceptable risks of failure? After all, less sophisticated weaponry can be less reliable for achieving a particular level of destruction. And you know that less firepower up front usually means that your forces will have to follow up to achieve your desired level of destruction—thereby raising the possibility of more losses and delayed task completion.

Eighteen hours elapse. You've been monitoring feedback regarding battle damage, tracking the progress of your three tasks. Your new task seems to be going surprisingly well. Your forces have met with little resistance, and all your weapons have found their mark. The weather has definitely helped the odds turn to your favor, even more than you had anticipated. If this continues, you believe, success seems practically guaranteed. But you know all too well that luck is a notoriously fickle ally.

Unfortunately, your other two tasks haven't faired as well. Three platforms have been lost, and few munitions have hit their mark. You tell yourself that it's always possible those three platforms wouldn't have been lost if you hadn't been using so many of your defensive assets in your new task, hoping to reduce the chance of losing a platform there. Further, it seems to you that these other two tasks are falling behind, and you don't have the extra assets to give them to get them back on track. Or do you?

All told, the question you really face is whether all your tasks are feasible. Nevertheless, the more fundamental question concerns what, if anything can be done now for success in the future. Should you take the risk of moving some of your more sophisticated weaponry back to your other tasks? Would that make the successful prosecution of all three tasks more likely? Perhaps you don't need so many defensive assets to finish your new task. Would moving some of those assets back to your other tasks help you maintain acceptable levels of risk for all three? If not, then you need to inform your superiors that success, as they've defined it, may be in jeopardy.

Are you ready to take that course, convinced you've examined all reasonable options and have a substantive, quantifiable basis for your conclusions?

You freely admit it: what you need is a tool that can help you manage the uncertainty of combat, so that you can make rational, rigorous, quantifiable decisions about what your next steps should be. Such a tool would be able to analyze the uncertainty that stems from interactions between your assets and the enemy's. Then, it would project that analysis into the future, to the end of your task, so you can see whether your task's success is in jeopardy. Better yet, if you are to meet your task objectives, you'd prefer that tool to make recommendations as to what your next step might be and, based on battle damage feedback, modify those recommendations. Finally, and perhaps most important, given the resources at your disposal, you would want the tool to help you rigorously assess the risks associated with your asset employment alternatives so that you could determine whether your tasks are feasible.

Indeed, this is just the kind of tool that researchers at Honeywell Laboratories, Minneapolis, Minnesota, along with subcontractors at the University of Southern California, Los Angeles, California, have been developing.

HELP ON THE WAY: THE TASK MANAGER

Answering the fundamental question of what it will take to get a task done, Honeywell's Task Manager makes recommendations regarding what's required now for success in the future. Moreover, Task Manager provides answers to this question not merely at the beginning of a task but continuously as the task unfolds over time [1].

Further, Task Manager exploits feedback about battle damage, so that as the battle progresses the risk associated with decisions about asset employment can be continuously managed. The tool thus helps military operators control the uncertainty that lies between current decisions about asset employment and the ultimate outcome of a task in terms of deadline, probability of success, and friendly-force losses. Task Manager also provides information about the resources required for maintaining the level of risk defined by task objectives.

To perform all its functions, Task Manager uses predictive models of battle dynamics to generate Minimal Effective Force (MEF) options. Each MEF option specifies the minimum amount and mix—or, size—of offensive and defensive assets required for the next combat event in a task. Further, each MEF option is designed to maintain the desired level of risk for task success and friendly-force losses over the time horizon implied by the task deadline.

Importantly, MEF options are designed to meet the acceptable level of risk from a task perspective, rather than from a short-term perspective that focuses on the likelihood of success for each combat event within a task. As used in this chapter, task means a sequence of engagement events that may extend for as long as several days and will involve multiple assets and multiple missions. The term task also implies that the task execution will include the continuous collection of information for battle damage assessment and utilization of feedback based on this assessment. This task perspective means that every MEF option is generated first, by considering the probabilistic distribution of future states (that is, force forecasts based on possible combat event outcomes) that would arise if this option were implemented; then, by projecting the MEF options needed for subsequent combat events, given the outcomes of prior events. Thus, each MEF option is built by considering all the chains of possible outcomes and subsequent MEF options—until the task is complete.

Military operators might use Task Manager as follows. After a broader, strategic objective—capture city C, for example—has been established, higher authority translates the objective into an interconnected web of individual tasks. Higher authority then conveys these individual tasks to Task Manager by placing such statements as "destroy designated target array by deadline D, with a probability of success S, and no more than L losses to friendly force" into a queue. The Task Manager, in turn, uses such task specifications to generate MEF options for each task.

Along with the general task specification, the Task Manager operator will input a description of the types of assets believed to be required for accomplishing the task—jammers, for example, bombers, fighter escorts, or other specialized types of platforms and weaponry. The operator will not input the quantity required, because Task Manager will calculate that. The final input required by Task Manager is a description of enemy assets—and this input function, like that of task specification, could easily be automated.

Task Manager then computes MEF options for all combat events in the task. Because Task Manager's computations are based on analytical solutions, as opposed to Monte Carlo methods, it's fast. Indeed, if Task Manager's analytical calculations were to be performed by Monte Carlo simulation, thousands of runs would be required just to generate the statistics needed for analysis. In the end, Task Manager will scale up to larger problems than those used in the current research, and therefore can be more readily used in real-world military operations as a battle progresses.

Each MEF option specifies the minimum size of each asset required for the first, or next, combat event in the task, plus the size of assets required for any subsequent combat events in the task. Furthermore, because there is typically more than one way to accomplish a combat goal, multiple MEF options are generated for each combat event. Here, each MEF option is a solution for the task as defined—meaning that all options are equally likely to lead to a successful task outcome, as measured by task deadline, probability of success, and maximum acceptable losses. MEF options for a particular combat event differ in the ratio of offensive and defensive assets required in the force, and thus provide military operators with the flexibility to choose the option that best fits current situational demands.

Task Manager uses feedback about battle damage to update and fine-tune MEF options for subsequent combat events. As the task unfolds, and if feedback is available, the process of calculating MEF options is repeated. Thus, feedback allows Task Manager to correct for both the inherent uncertainty of combat and the inevitable inaccuracy in intelligence information regarding enemy assets.

FLIPPING A COIN

To military operations, Task Manager applies model predictive control (MPC), a paradigm successfully used for the last two decades in such industrial applications as control of large and complex manufacturing and chemical plants. Unlike industry, however, the military has to contend with an intelligent adversary bent on disrupting operations. Assets on both sides are lost and gained— all with a significant degree of uncertainty. Thus, Task Manager must include such considerations in applying the industry-proven paradigm of MPC.

As the name of the paradigm implies, Task Manager uses models, predictions, and controls. The models concern battle dynamics and the uncertainty associated with the outcome of combat events. More specifically, Task Manager's predictive models of battle dynamics include kill probabilities for interactions between different types of friendly and enemy assets (artillery and bombers, for example, or bombers and bridges) in both directions during a combat event. Such models of battle dynamics provide a mechanism for using the current state of the battle to predict (probabilistically) its future states. Here, such predictions concern how decisions about assets to be used for the next combat event, along with the implied asset requirements for subsequent combat events, will impact the likelihood of task success. The controls concern the number of assets that should be employed in the next combat event for eventual task success.

This book's Chapter 2 describes key concepts of MPC. Using that chapter's terms, Task Manager is a closed-loop controller that performs stochastic optimization for a time horizon with a fixed end-point; that is, the horizon gets progressively shorter as the task deadline approaches. Key points are that Task Manager is aware of the impact its recommendations may have on operations, and thus generates MEF options in a self-referential manner. Task Manager thus uses end-point reasoning to trade short-term gains for long-term gains, and can use feedback during battle to make corrections prior to that end-point. Finally, it works with probabilistic outcomes to make predictions.

The computation performed by Task Manager to generate MEF options can be described very roughly through analogy with a coin-flipping game. Imagine a game in which the player flips a number of regular coins for a fixed number of rounds. After each round, the player scores a point for each coin that comes up heads, but loses the coin if it comes up tails (and so the coin cannot be flipped in the following round).

Heads or Tails?

In this analogy, an example of a task is to score seven points with a 50 percent probability of success in a deadline of three rounds. A corresponding military task might have seven targets in a target array that must be destroyed. Thus, scoring points is analogous to target destruction, while the coins are analogous to a friendly asset with a 50 percent probability of kill as it interacts with each target in the array. Here, then, losing a coin corresponds to the 50 percent probability of the target destroying the friendly asset. A round in this game is therefore analogous to one combat event. Since the deadline in a corresponding military task would probably be specified as a time by which the task must be completed (for example, 1800 hours tomorrow), the deadline can be converted to rounds in this analogy with a speed/distance calculation. Thus, in the corresponding military task, the speed of the asset and distance to the target are such that a maximum of three ingress/egress events could be accomplished in the time allotted. Together, target destruction (that is, scoring points) and losing friendly assets (losing coins) comprise the possible outcomes of a combat event. These outcomes accumulate over the time course of the task as combat events are successively executed.

If the game were only for one round, then it is intuitively apparent that one should flip 14 coins (that is, use 14 friendly assets) to achieve seven heads with a 50 percent likelihood. Hence, the solution is 14. As the number of rounds increases, the calculation becomes more complex and less intuitive. Since the outcome of any particular round is uncertain, Task Manager uses an internal model to predict what will happen.

More Than Chance

The question of interest for Task Manager is what is the minimum number of coins to flip in the first round to guarantee a 50 percent chance of scoring seven points in three rounds. It turns out, not entirely intuitively, that the rigorous solution is eight. The logic behind this solution can be illustrated by tracing a single sample path through the possible states (or, outcomes of each round) in the game. If eight coins are flipped in round one, you can expect to get four points and lose four coins. Flipping four coins in round two should give you two more points and a loss of two more coins. Finally, in round three you can expect one more point (and one more loss) for a total of seven points. Of course, one has to average all the possible outcomes to get the right solution, but the answer, eight, is akin to the MEF options that operators get back from

Task Manager. Note, however, that just one of many, many possible paths has been described; it is these multiple possibilities that make the problem both difficult and hard to manage intuitively.

By evolving the dynamics of the internal model with a choice of eight initial coins, it can be shown that there is a 54 percent chance that seven heads will be scored in three rounds. The key point here is that eight is the minimum required to score seven points with 50 percent confidence. A choice of seven coins initially gives only 42 percent confidence. More coins would of course increase this percentage.

From the coin flipping analogy, it can be seen that Task Manager doesn't tell military professionals how to make decisions about force employment. Instead, Task Manager gives them ways to evaluate the degree of risk that their decisions imply. Fewer than eight coins in round one would jeopardize task success. Similarly, more than eight coins in round one would jeopardize task success—because the goal was to flip seven heads in three rounds, not three rounds or less. Thus, using too many coins (or resources) is ultimately no better than using too few.

SCARCE RESOURCES

Now it may be argued that using extra resources, and therefore finishing a task early, is acceptable—indeed, perhaps preferred. Although such acceptability may be true in some cases, in most cases it isn't—because such a judgment requires consideration of the bigger operational picture, one that involves other tasks. In a resource-rich environment, using extra assets is no problem. But if resources are scarce, using extra resources for one task means taking resources from another, which may increase the risk of the second task failing. Further, the use of excessive resources may, under certain circumstances, lead to a greater likelihood of friendly losses.

Finishing early may also be fine—if you're not concerned with the reasons why the person who issued the task specified that particular deadline. Indeed, in all likelihood the task-issuer will have very good reasons for specifying a particular deadline. Otherwise, why bother? Such reasons will most likely concern the coordination of multiple tasks, so that the timing of your task's completion impacts the likelihood of other tasks' success or failure. For example, the deadline specified for your task may have been determined by estimating how quickly the enemy can repair the damage from your task—given that subsequent tasks must be executed before the damage is repaired.

The use of feedback is an essential and integral part of Task Manager. In the coin flipping analogy, feedback could be incorporated by allowing the player to add or remove coins (resources) at the end of each round, depending on the outcome of the coin tosses. For instance, if the player were lucky and flipped six heads in the first round, then—as only one more point needs to be scored—in the next round there would be no reason to flip all six remaining coins. In this case, the player would be better off flipping only two coins, thus avoiding risking the other four coins, which in turn could be used for a different game. Feedback, therefore, helps fine-tune the management of risk, which, along with the use of MEF options, helps to free resources for other tasks.

OVERCOMING COMPLEXITY

It's easy to see how quickly modeling battle dynamics can get complex. In essence, the coin-flipping analogy reflects a situation in which there's only one type of asset with a probability of kill (and being killed) of 50 percent. Furthermore, during each round, the coins don't get bent or otherwise altered, which would change the probability of heads. In addition, no strategy exists to alter the probability of particular coins killing or being killed. Thus, modeling battle dynamics is decidedly more complicated than implied by the analogy [2].

A key benefit to be gained from Task Manager is that, by proposing MEF options whose composition is based on rigorously calculated risk and therefore can be rationally argued and justified, Task Manager enhances the objectivity of task planning and asset allocation. By accounting for available feedback, another key Task Manager benefit is the ability to control the task continuously.

Using Task Manager is straightforward. You enter your task objectives, the assets you believe are needed, and the assets you believe the enemy has. Task Manager, in turn, returns a set of MEF options—quantities and mix of assets with different capabilities. Then, based on your experience, you can look at those options to see if they're reasonable, selecting the one that seems most appropriate for the next combat event. When battle damage feedback comes in, you use Task Manager to update MEF options for subsequent combat events. This transparency is part of the elegance of the Task Manager approach.

With Task Manager, you can define tasks in familiar terms: at the least, all tasks have a deadline, a required probability of success (either in quantitative or qualitative terms), and a statement about acceptable losses. If necessary, other performance objectives could be added to extend Task Manager—with the only imposition that Task Manager requires objectives be stated explicitly. Whether these objectives are handed down by higher authority, or determined through what-if analyses with Task Manager, depends on the particular concept of operations chosen.

To elaborate, task deadlines provide a mechanism by which, as part of a bigger operational picture, task issuers can coordinate multiple tasks. An example is in this chapter's introduction, where a path had to be cleared by 1800 hours for an upcoming offensive—another task. Today, probability of success is typically not talked about, at least not explicitly. Even though it is understood that in reality task completion can never be guaranteed, because of the inherent uncertainty of combat, nevertheless, as military operators select a COA, they must consider the odds of success. In practice, Task Manager's requirements for the probability of success both reflect the importance of task accomplishment and specify the assurance the task issuer requires for completion. In terms of acceptable costs to the task issuer, the parameter for maximum acceptable losses reflects another measure of task importance. As assets may be required for subsequent tasks, this measure also has to be determined by considerations related to the bigger operational picture. As in any military campaign, political sensitivities may also be at issue here.

When issuing a task, stating a required probability of success of .95 is equivalent to saying that 95 percent of the possible paths through the space of all combat events must lead to a successful task outcome. Thus, each MEF option must lead to task success 95% of the time, by the specified deadline, and with losses that don't exceed the maximum allowed. Importantly, the probability of success is more than the probability of target destruction. In the Task Manager framework, successful target destruction is a necessary, but not sufficient, condition for successful task completion. Targets to be destroyed represent only one dimension of the task objective. Indeed, tasks also have other important dimensions, including deadline and friendly loss. Even if the entire target array is successfully destroyed—but with unacceptably high friendly losses, or after the deadline—the Task Manager views the task as a failure. In this light, then, the task specification is a multidimensional objective, defining the target array, deadline, friendly loss, and possibly other aspects. Thus, Task Manager helps operators manage risk on three dimensions—or more, if necessary.

DIVISION OF LABOR

Task Manager clearly separates definition of objectives from details of planning and execution. This clear separation, in turn, provides for the possibility of a concept of operations where those who define tasks can be shielded, if desired, from the details of resource management, planning, and execution. In theory, tasks can be defined without regard to either required resources or any details associated with how the tasks might be executed. Thus, Task Manager supports a concept of operations in which operators can first decide what needs to be done and then determine what resources will be required.

In practice, it would not be prudent to define a task with complete disregard for the resources it might require. Generally, if task issuers want to decrease risk or complete tasks quickly, they must be willing to pay the price in resources. All else being equal, resource demands grow as acceptable risk decreases. For an extreme example, because of the inherent uncertainty of combat, certain success—a required probability of success of 1.0—would require infinite resources. By the same token, shorter time horizons increase resource demands.

If acceptable risk or deadlines are relatively fixed, Task Manager can be used proactively to determine what resources should be brought into the theater for upcoming tasks, thus providing an analytical rationale for needed resources. However, if resources are relatively fixed but deadlines and risk are more flexible, then, given the resources available, Task Manager can be used to assess tradeoffs among deadlines and risk. Either way, Task Manager provides information about task feasibility in terms of MEF required. Furthermore, by adding battle damage feedback, and examining updated MEF options, task feasibility can be continuously tracked.

A great deal of Task Manager's transparency stems from how its predictive models of battle dynamics relate to the combat domain. These models, in effect, are used to predict future states of assets currently involved in the task. Such predictions therefore require that different models be used for different types of combat events— or, more specifically, for combat events that require interactions among different types of assets. When operators specify friendly and enemy assets involved, the operators are in fact selecting an appropriate model of battle dynamics.

To examine the probabilistic outcomes of asset interactions, projected over the task's time course, Task Manager consults the appropriate model to output a set of MEF options, which, regardless of the specific model used, are generated in the same manner. Thus, simply by changing the underlying models of battle dynamics, MEF options can be generated for very different types of tasks, involving very different types of assets. The use of Task Manager does not change.

Such a course raises the possibility of a concept of operations where Task Managers are distributed among operators responsible for very different types of tasks. If all operators report MEF options to a centralized location, then Task Manager in effect provides an integration capability, whereby higher authority can view the landscape of resource demands implied by bigger operational objectives. Here, the nice thing is that this landscape can be assessed without getting into any of the details of the underlying models used to compute MEF options. Closely related to this concept of operations is one where, as the type of task changes, or as new assets with new capabilities appear on the scene, a single operator swaps, or otherwise modifies, models used by Task Manager.

The strategies used on both sides during combat events are also likely to be an important influence on the lethality of interactions between friendly and enemy assets. If the models used to control risk are to portray the reality of the situation accurately, then they must be modified. If, for example, doctrine and other intelligence sources indicate that the enemy prefers targeting one type of asset over another, then any combat event using that asset is exposing it to a higher probability of being killed than under a different strategy. One's own strategies for the conduct of a combat event will also alter the lethality of interactions.

AUTOMATING INPUT

Strategy on both sides will also impact the model that Task Manager uses to predict both combat outcomes and eventual task success. Strictly speaking, then, the military operator must also input strategy so that appropriate models of battle dynamics can be selected—or, if they are not available, constructed online. To this end, Task Manager researchers both envision automating a large portion of this online model adaptation process, and have worked on defining elementary combat models, the beginnings of user-interface requirements, and knowledge-based support requirements to assist operators [3]. On a somewhat different note, Task Manager's probabilistic models of battle dynamics offer several important advantages over classical Lanchester-like, attrition-based deterministic models. By explicitly accounting for the randomness always present in combat, the Task Manager models provide a means for evaluating variability in the expected outcomes of combat events. Such a focus on variability allows Task Manager to shift from attrition optimization to risk management [4].

Task Manager's shift to a risk-management framework eliminates the need to communicate asset valuation across the C^2 hierarchy. Perhaps one of the most important benefits of Task Manager is that its MEF computations don't require any notion of target value, or more generally, asset value. Thus, Task Manager technology offers a way to solve the perennial problem of determining what resources are needed for a task without requiring military operators to deal with the frequent—and largely unsolved—problem of assigning values to enemy targets and friendly assets.

Asset valuation, which amounts to ranking friendly and enemy assets in terms of their relative importance, is difficult because an asset's importance will depend on how it fits in the bigger operational picture. Consequently, asset valuation will be situation-specific and, quite likely, time-dependent. The value of the precision-guided munitions in this chapter's introduction, for example, would probably differ drastically before and after the receipt of the new task.

Using Task Manager to make resource allocation tradeoffs, operators, who may not have all relevant information to value assets rationally, are not required to value assets. Here, instead of requiring that operators know why and how valuable it is to destroy a particular target, or to protect a particular asset from loss, the task issuer conveys value implicitly, yet rigorously and quantitatively, specifying the desired probability of task success, along with maximum acceptable friendly force losses.

With Task Manager, as elsewhere, no battle dynamics model will ever capture perfectly all the relevant information needed for force-employment decisions. Situation-specific considerations will almost always lead operators to prefer one force-employment option to others—a fact recognized by the design philosophy behind Task Manager.

Accordingly, Task Manager sticks to managing the risk associated with the essentials, and provides the military operator with options that both meet minimum requirements and reflect that most tasks can be successfully accomplished in a number of ways. Since all MEF options are equal in their likelihood of success, operators may select one MEF option over another for any reason. Thus, without worrying about severe variations in risk, operators can use experience and expertise, along with their richer knowledge of the situation at hand, to select among MEF options.

EXTENSION TO A BROADER FRAMEWORK

Task Manager also offers a framework, as seen in Figure 8-1, that allows for a building-block approach to C^2 system design. First, models of battle dynamics form the core of the framework. Then, as described, Task Manager uses these models to compute MEF options—which in turn facilitates resource-allocation decisions.

Thus, Task Manager researchers built a Resource Allocator that, using MEF options for multiple tasks, first examines the landscape of resource demands, and then selects the first resource allocation scheme that meets all MEF requirements [5].

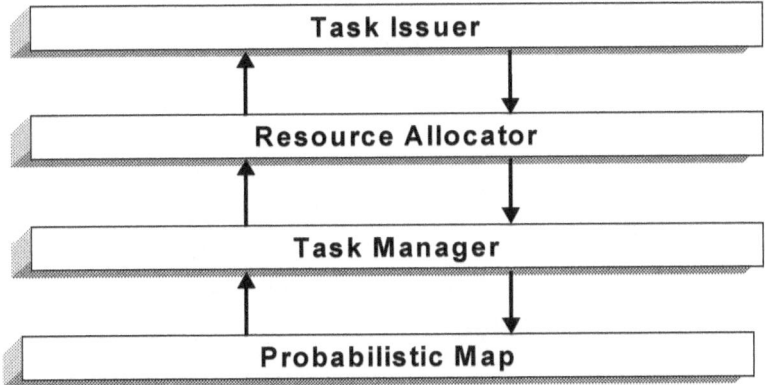

Figure 8-1. The Task Issuer, Resource Allocator, Task Manager, and Probabilistic Map components encompass the Task Manager framework that supports the building-block approach to C² design.

Once it's known that resource supply meets resource demand, the time has come for more detailed combat event planning. Thus, researchers have built a Probabilistic Map technology that, by managing risks associated with the spatial configuration of enemy assets, helps operators during route planning. At this more detailed level of reasoning (that is, in terms of probabilistic locations of enemy assets), risk assessment can, in turn, be used by Task Manager to refine expected lethality, itself used in the computation of MEF options [6]. Superficially, the division of labor and reasoning in Task Manager's framework remains hierarchical. In its functionality, however, there are substantial differences in the way information flows about the hierarchy. The strong, top-down information flow traditionally associated with hierarchies is complemented by a comparably strong bottom-up flow. The output of Probabilistic Maps, for example, influences the higher-level computations of the Task Manager, which in turn influences the Resource Allocator's decisions.

With the objective to match optimally the broader set of military objectives (or, multiple task objectives) with the resources and operational constraints existing in a battlespace at any given moment, Task Manager information flow produces a structure in which such decisions are reached through a negotiation process.

Combating Unfeasibility

When task unfeasibility occurs, it can trigger an extensive negotiating process that runs up and down the reasoning hierarchy. So that tradeoffs among objectives can be assessed, the person who issued the task can ask for sensitivity analysis of other tasks in progress, along with their status and prospects. In order to make room for new tasks, and taking into consideration lower-level, battlespace-related constraints, some tasks can then be modified or dropped.

With Task Manager, such negotiations can happen automatically. Although in the near-term, reasoning at these different levels requires a combination of human controllers and risk management technology; such roles may be fully automated in the future—if desired. Indeed, as more and more unmanned fighting platforms come onto the scene, automation will become increasingly acceptable. With Task Manager, because all negotiations can occur in machine time with only high-level supervision by human specialists, when a large degree of automation is in place the framework allows for very fast reaction to battlespace changes.

Warfighters at the Round Table

The gist of the Task Manager framework can be portrayed as a team of technology-assisted warfighters seated at a round table. While each team member's responsibilities are clearly divided per the described reasoning hierarchy, through negotiation they work together to achieve overall operational objectives. Such a clear division of labor, and the rational negotiation among team members, is supported by a seamless integration of various technologies within Task Manager's MPC framework. Here, co-location is not a requirement: indeed, one of the important benefits of Task Manager's framework is that this risk management technology can be distributed among geographically dispersed operators, thus removing the possibility of a single point of failure.

Equally important, technologies can be added, removed, or replaced without requiring major modifications of those already in the framework, as seen in Figure 8-2. So, for example, the landscape of resource demands output by Resource Allocator could be used as a basis for placing bids for new tasks via Auctioneer, as described in Chapter 10. Similarly, the output of the Probabilistic Map could be sent to lower level controllers, like those described in Chapters 13 or 15, to guide more detailed mission planning and execution, quite possibly in real time.

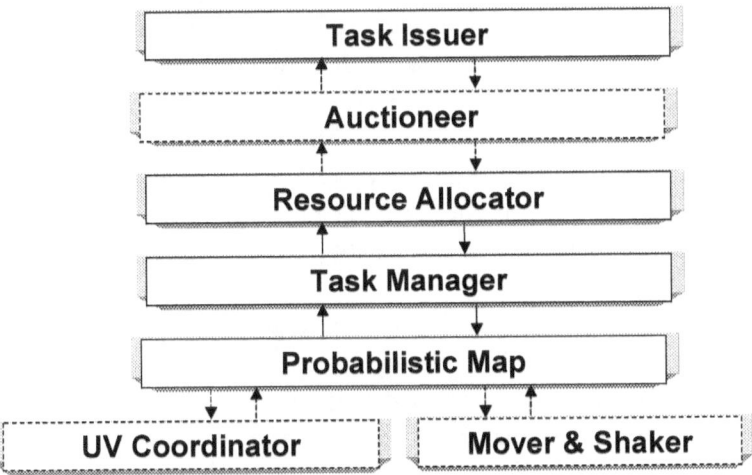

Figure 8-2. The Task Manager framework can add, remove, or replace components without requiring major modifications to those already in the framework.

DIFFERENT EMPHASES

From an operational perspective, Task Manager—and the larger framework that encompasses it—is not mature. Transition to a fielded system will require that necessary support for the technology, and the specialists who work with it, be specified and, if not already in place, developed. Some of the missing pieces are areas that require additional research. Others require technologies in the larger C^2 infrastructure that support Task Manager technology.

Doctrinal thinking may present Task Manager's biggest challenge; or perhaps attitudinal change is a better way to describe it. As a technology, Task Manager computes MEF. However, the mere utterance of that term often causes both a negative reaction and accompanying arguments about the great importance of possessing, or using, overwhelming force. The emphasis in the argument against Task Manager is usually on the word minimal, whereas the key counterpoint places emphasis on the word effective. Task Manager's computations statistically guarantee task-success based on the risk specified as acceptable by higher authority. As an example, if acceptable risk is low, its force recommendation will be indistinguishable from a force designed to overwhelm. Task Manager's force recommendations will therefore always be minimal—so that resources can be utilized most efficiently. Here, both minimal and effective are important for balancing resource demands and coordination among multiple tasks in the larger operational picture.

Other challenges to Task Manager will arise when humans consider task definition and distribution. Indeed, given the development effort focus thus far—on both the predictive modeling of battle dynamics and the mathematical intricacies of computing MEF options—such challenges should be expected. Here, the major challenge will center on decomposing higher-level objectives into sets of more specific, quantifiable tasks. Currently, it is unclear what information and domain knowledge will be needed both to develop and prioritize tasks, and to assign the required level of risk for task completion.

Concerns about coordination among tasks have also been raised. At this point in development, task coordination is seen as an issue that should be resolved by those task issuers who have access to the larger operational context. Giving each task an appropriate deadline will enforce, at least in part, task coordination. Likewise, task issuers must anticipate contingencies as well as make provisions for them. For each task, therefore, the acceptable risk of failure must be proactively managed by establishing the appropriate desired probability of success. Accordingly, coordination would in all likelihood take place through vertical channels—as opposed to lateral channels between operators associated with different Task Managers—because such a vertical path would reduce beneficially the potential for information overload caused by lateral communications.

With regard to the intelligence community, because Task Manager calls for a relatively rapid turnaround in the collection, processing, and dissemination of battle damage assessments, the technology seems to imply rather heavy intelligence demands. In general, although it is highly desirable to receive battle damage assessment after each combat event, such assessments are not mandatory for Task Manager's successful operation. Indeed, Task Manager updates its force calculations only after receiving new intelligence data, not necessarily after each combat event. Since Task Manager always recalculates all future combat events up to task completion, and uses the force calculated earlier to execute the next combat event, failure to receive an intelligence update is not fatal. Lacking the update, this execution is still the best option available. Moreover, if the original intelligence was of high quality, the occasional lack of new, up-to-date information will not significantly decrease the probability of success.

Finally, as parameters in its models of battle dynamics, Task Manager requires access to a lethality database. Here, given all the situation-specific variables that may impact accurate estimates (weather, for example, or strategy, day versus night operations, and so on), the challenge lies in getting such estimates of probabilistic interactions between friendly and enemy assets. In this regard, Task Manager researchers have demonstrated that its calculations are quite robust regarding mismatches between model parameters and true lethality numbers. In addition, based on battle damage to friendly forces, the research team has also implemented adaptive model predictive controllers that estimate enemy lethality and number of forces. Such estimated parameters are used to update the internal model of battle dynamics, which is then used to develop MEF options for the next combat event. In that this adaptive control approaches the performance of control based on complete and correct information, the results to date are promising.

SUMMARY

A tool like Task Manager offers a commander the means to manage task risks, both in planning and execution, in a rigorous, quantifiable manner. As defined here, a task is a sequence of mission and engagements, which must be accomplished by a specified deadline and with a specific probability of success. Many possible scenarios may evolve in the execution of such a task; a coin-flipping analogy demonstrates how unintuitive the solution can be.

Before executing a task, a commander uses Task Manager to predict the composition and size of the force necessary to accomplish the task within specified parameters. The tool takes into account the uncertain nature of future outcomes, in effect considering the entire tree of possible future events. As the execution of the task unfolds, Task Manager, using available feedback about events that have occurred, fine-tunes the recommended options for forces to employ in executing subsequent combat events.

The chapter also introduces the concept of Minimal Effective Force—the minimal force required for executing a task, taking into account uncertainty and specified risk constraints. The emphasis is on effective—on helping the commander explore what is sufficient and what would be too risky—while leaving the decision to the commander.

In control terms, Task Manager is a closed-loop, model-predictive controller that performs stochastic optimization for a time horizon with fixed end-point. It includes probabilistic models of engagements, and explicitly, analytically calculates solutions for the required task. Because Task Manager does not use expensive Monte Carlo simulations, it can work fast, providing help in real time, and can scale up to very large problems. The approach accounts not only for uncertainty in combat, but also for the inevitable inaccuracy and incompleteness of intelligence information.

CHAPTER 9

HIERARCHICAL NEGOTIATION
AND DELEGATION

A complete C^2 system contains many interrelated components. Could there be a system that would integrate all those components? To borrow an analogy from anatomy, the system would have to be built like a human, with a skeleton (a sound and theoretically rigorous hierarchical architecture), muscle (models and optimizing algorithms well-suited to exploiting the underlying architecture), and skin (a friendly interface tuned to military ways of doing business).

To illustrate the capabilities of this technology, consider an Operations Other Than War (OOTW) scenario; a relief operation following a devastating Central American earthquake. Plausible operational experiences using HELPS (Hierarchical Expeditionary Logistical Planner and Strategist), a notional C^2 support system based on the distributed control prototype created by a team headed by the Charles Stark Draper Laboratory, Inc., Cambridge, Massachusetts, with support from Science Applications International Corporation, and ALPHATECH, Inc., Burlington, Massachusetts, will be described. The Draper hierarchical design achieves enterprise-wide optimization while delegating decisions to the lowest level that can solve the problem. This will be contrasted with experience using a hierarchical C^2 support system that attempts global optimization by centralizing much of the decision making. Through this OOTW scenario, the design advantages of the proposed technology will become clear, while further sections describe it in greater detail, discuss implementation issues, and identify remaining technical challenges.

AN OOTW MISSION

In the first few days following a devastating Central American earthquake, an OOTW rescue and relief mission is launched. A key issue on the mind of the Commander, Joint Task Force (CJTF) is the balanced disposition of the resources—human and hardware—under her command. Some resources will be scarce and valuable (helicopters, say); others will be more plentiful (ambulances), but perhaps with somewhat limited capabilities. She also knows that she can expect plenty of surprises, for example, the discovery of casualties in immediate need of medical assistance, unanticipated road blockages and power outages, and so on. Finally, human lives are at stake, and the press is on hand, so effective, prioritized use of distributed resources is paramount. At the outset, if she could articulate what she would most want from the C^2 support system, it might be that the system not interfere with operations. First, do no harm.

The issues at stake fall into four vignettes encapsulating encounters the CJTF might have over the mission's first 72 hours with HELPS, the C^2 support system. Hierarchical in nature, HELPS combines mathematical intelligence with practical military experience—a virtually unbeatable combination.

The Staffing Report

Predictably, the first 24 hours are hectic. First, the JTF components are put on alert. Following confirmation from the National Command Authority that an official request for assistance has been received, JTF units begin to deploy to their assigned destinations. On the flight down, an aide brings the CJTF a digital tablet showing a map of the current intelligence on damage, hospitals, bases, and so on. Using a pen, the

Commander circles two regions where heavy damage is blocking highway access. "We have to get these cleared," she says, writing the letter H (for High Importance) in the circle, "so that we can start using ambulances to evacuate these three villages." Again, she points, marking the features of interest.

After 20 minutes with the digital tablet, a complete set of priorities—geographic regions, asset types, mission objectives, and associated time deadlines—has been entered. Concurrently, HELPS software is mapping these semiotic entries into the numerical form needed by the underlying algorithms. As the mission progresses, entering new Commander's intent is just as simple: erase and redraw to suit current conditions.

The Commander recalls working with an earlier prototype C^2 system several years ago, one that required numeric entries and values for each entity (both Blue and Red) in the battlespace, as well as every objective and task in the plan. Then, the system needed human translation and understanding of the mathematical algorithms, resulting in the addition of an O6 position titled Value Coordinator and two support personnel to the JTF staff. Thankfully, the HELPS system interface has no numbers, no translation and no additional staff—just a natural and effective way for the Commander to interact directly with the system.

The Ambulance and The Helicopter

A report comes in that Armando Castille, ambassador of a friendly nation, has been trapped by falling debris and needs immediate assistance. Real-time replanning begins, and the Ambassador is now identified as a Very High Priority mission objective. With resource decisions kept at a local level, personnel at the ambulance staging area know that the bridge connecting their base to Embassy Row had just been repaired—greatly reducing transit time. The HELPS system distributes and delegates decision making to those levels that have the best and most immediate information and the most relevant expertise. Because such information may not have percolated up the hierarchy, it would not be available in a global optimizer. Even more significant, however, is that because meeting time constraints requires compromising model fidelity, models used in global optimization are often low-fidelity. Hence, tradeoff decisions are often based on large-scale generalities as opposed to fine-grained, detailed analysis.

All these gradations of intelligence would be of only minor significance, except that within minutes, a second—and far more serious—emergency arises, involving school children trapped by a natural gas leak explosion. In retrospect, if a helicopter had been sent to rescue the Ambassador, instead of an ambulance, CNN could have headlined: "Ambassador's life preferred over children's."

The inadvertent public-relations disaster did not occur because HELPS works hard to keep decisions as low in the hierarchy as possible. This outlook has several benefits: the potential for more accurate and timely sensor inputs; better fidelity models; less disruption to other units; and, under certain circumstances, the ability to apply scarce resources more effectively for as-yet-unforeseen eventualities.

Second, HELPS tempers priority assignments by adding a planning inertia that both prevents wild planning swings and allows time for confirmation and weighing of alternatives.

Third, by employing a sophisticated optimization algorithm exactly tuned to its hierarchical architecture, HELPS permits high-fidelity models throughout the system. By automatically decomposing the global problem into sub-problems that can be solved at lower levels in the hierarchy, HELPS instinctively pursues solutions at levels where more detailed information and models reside. In solving this global optimization problem, then, HELPS would not have sent the helicopter, so that it would have been available when the school went up in flames.

Restlessness Among the Troops

By the morning of the third day, the large number of parallel activities reveals a coherent path to recovery because HELPS includes plan stability in its objective function. That is, in addition to asset and mission value, HELPS compares the degree of difference between the old plan and the new one at each event-triggered and periodic replanning cycle. That difference is then considered as part of the optimization scheme. Solutions that result in large and frequent changes are automatically penalized—and are only used when the operational benefits of the proposed changes are clear for all to see. Here, then, HELPS provides a mechanism to establish the proper balance between stability and agility, an important practical consideration noted in Chapter 3.

In contrast, the global optimizer would have resulted in a C^2-caused morale and planning mess with complaints about ever-changing priorities, tasks, and missions coming in regularly from lower echelons. The problem is related to the frequency with which previous C^2 systems would reoptimize, and the detrimental effect such reoptimization had on morale. In addition, re-optimization caused difficulties. First, since radical and frequent plan changes had little or no discernible payoff in terms of mission objectives, the C^2 system lost credibility. Second, because crews could not count on being able to use the time spent in mission preparation and rehearsal, morale was down. Indeed, plans changed two or three times before take-off-and again en route! Third, all the changes were stymieing logistics. As is well known, predictable routine has great military value in its own right, so it's advisable to temper putting the virtues of agility far ahead of stability.

Indeed, the Commander might not even have been aware of this as she okayed the stability-preferred default box in the HELPS start-up dialog; but she would certainly have preferred to avoid the complaints and loss of morale occasioned by an adherence to agility at all costs.

Too Much Control

Centralized C^2 can be problematic if all decisions within the hierarchy require human review and approval, especially if the review and approval is only at a higher echelon in the hierarchy. The issue here is the correct balance between local control and global optimization. As previously stated, with input from lower levels in the hierarchy HELPS negotiates its top-level solutions. Further, unless arising circumstances warrant otherwise, HELPS can also set triggers and thresholds that force decisions to stay at the tactical level. In short, HELPS has the common sense to balance all the factors that enter into an effective operational system. In the present scenario, even if she chose to monitor decisions, the Commander would not switch off HELPS. Why not? Because, HELPS does not overwhelm its human counterpart by elevating the simplest local matters to a global decision. (This problem is discussed generically in Chapter 3.)

With the earlier C^2 decision support system, it was decided to review top-level decisions manually after the helicopter/ambulance fiasco—a kind of quality control measure to insure that the decision passed a basic sanity-check. What nobody realized, however, was how many decisions were made at the top level of the hierarchy. Even seemingly simple local decisions would percolate upwards, often resulting (as we saw above) in reoptimization, and hence affecting large portions of the network. Instead of only a few top-level decisions to review, the operator assigned to monitor decisions is constantly under pressure—and the central command post becomes a bottleneck. Lower levels in the echelon need to make changes, but must wait for central approval: decisions are pushed as high up as possible because of the rationale that global optimum is better than local. Such a philosophy might work when human review is not required; but the addition of a human monitor introduces intolerable latencies, and the system is in danger of collapsing under its own weight. What could bring down an entire system? Something as prosaic—and seemingly inconsequential—as a flat tire. A scheduled ambulance is no longer available, and instead of simply working around the problem at the local base, a centralized C^2 system could decide to reoptimize the entire JTF plan.

The After-Action Report

The Commander's reaction to a HELPS-like system—that is, a hierarchical system that incorporates C^2 controller features—is very positive. Indeed, the after-action report probably wouldn't even mention HELPS at all. Instead, it would focus on matters that should occupy a commander's attention: the mission itself, and its accomplishment. After all, to be truly effective, a C^2 support system should be almost invisible, a component that is both taken for granted and is as integral to a warfighter's kit as an M2 or a walkie-talkie.

Does HELPS provide enhanced planning processes not found in current C^2 systems? Absolutely. Today's C^2 consist primarily of stove-piped systems that do not permit integrated planning and execution management between echelons. At the strategic level, the Joint Operations Planning and Execution System (JOPES) does not support end-to-end, time-phased, force-deployment data (TPFDD) development.

Tools within the system permit TPFDD building and testing, but there is no connection between the campaign plan and force-flow planning. As execution proceeds and the actual force flow deviates from the plan, there is no means for feedback into the planning system that will assess the disruption's impact on the Commander's plan. Indeed, the flow of logistics is modeled within the planning system for the purpose of estimating transportation requirements, but without the detail necessary to plan logistics requirements—

or logistics operations—accurately. The problem, then, is not going away. Over 10 years ago, a book explored this problem in depth [1]. After a decade of development, it is still reported that "planning processes and tools fail to support the requirement . . . [and] the supported CINC's needs are not satisfied by using the currently accepted methodology [2]."

A HIERARCHICAL C² ARCHITECTURE

The underlying technology that makes the notional HELPS system possible has its genesis in the more descriptive and accurate designation: Hierarchical Command and Control Architecture (HC2A).

The Basic Building Block: A C² Node

HC2A is composed of interconnected blocks, or modules, called C² nodes. Each may be thought of as a more or less independent collection of software entities with well-defined interfaces (data and control) to other C² nodes in a well-defined hierarchical relationship [3]. The structure formed this way is a tree, albeit inverted, with a root node (the top-level commander) at the top, intermediate nodes down the tree, and finally leaf nodes (the lowest-level tactical units) at the bottom. Thus, the HC2A block diagram corresponds to the hierarchical organizational structure of current military practice.

Except for the top-most and leaf nodes, any given node in HC2A will be connected by data and control communication links to exactly one node above (its superior), and to one or many nodes below—its subordinates, toward which it acts as the superior. By working up the structure, one node at a time, through a series of intermediaries even the lowest nodes can eventually be connected to the top-most command node. This structure should be both familiar and intuitive, for this structure is nothing but the standard hierarchical organization chart, with worker bees at the bottom and the brass at the top!

HC2A focuses on what is the same at every node—that is, what basic functions must be performed no matter where in this hierarchy a node is located. Yes, the details of that functionality within each node will differ; nevertheless, no matter where it is in the hierarchy, there is a commonality of function across all nodes that can be abstracted as the basis for node construction. A way to think of this construction is as a template which stays the same for all nodes, and which is then colored or customized for the particular function any given node must perform. The idea is then to replicate the structure and customize the details.

Figure 9-1 shows a generic block diagram of one C^2 node, including its up and down interfaces. Four functions are shown: monitoring (M), diagnosis (D), planning (P), and execution (E). The monitoring function compares received state information (much like extra-strength military intelligence reports) against predicted state information, and determines whether the situation has changed enough to require replanning.

In most cases, it is hoped—and expected—that the monitoring function need not take further action. This is an extremely important notion: most of the time, system response is routine and can occur at a very high tempo. So long as the residuals (predicted states minus observed states) are within reasonable bounds, the system—›monitoring—›execution—›system (S—›M—›E—›S or simply SMES) loop can continue at very high rates. The execution unit, in turn, may be thought of as the currently selected control law, accepting state estimates from the monitor as input, and generating control signals either to battlespace entities (at the lowest, tactical level) or to subordinates (at intermediate and higher levels of the hierarchy). Conceptually, each node thinks of whatever is below it in the hierarchy as the system to be controlled.

Should the monitor detect that the system state is straying outside of acceptable limits, or that an opportunity has arisen, the diagnosis function is notified to take corrective action. Its first recourse is to try a completely local solution. That is, it may be that the C^2 node can adjust to the perceived change simply by selecting a new control law (hybrid controller techniques), adjusting system parameters (adaptive controller techniques), or reoptimizing (negotiation with subordinates). In any case, it is hoped that most exceptions can be dealt with using the S—›M—›D—›P—›E—›S loop without the need to notify a higher supervisory level in the hierarchy. The time needed to use this loop will be longer than if the autonomic SMES loop were taken, but (most important for our purposes), the upper levels in the hierarchy have been shielded from unnecessary detail. Thus, the S—›M—›D—›P—›E—›S loop is fallback protection against excessive centralization.

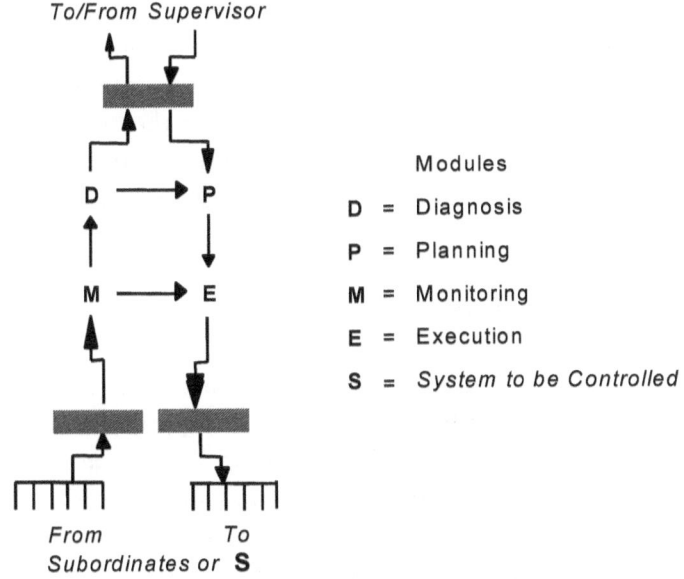

Figure 9-1. A generic C^2 node shows four modules: Monitoring (M), Diagnosis (D), Planning (P), and Execution (E).

In a good design, the invocation of the planning function will result in a reset of the monitoring function as well. While this interface is not shown in Figure 9-1, it is essential, since the hope is that, with a revised prediction and control law, the system will return to autonomic SMES behavior.

It may occur, however, that the diagnosis function determines that the problem is so severe that corrective action will require the intervention of a superior in the hierarchy. In that case, the information is passed to the next-highest level. The superior is provided with a properly summarized set of state estimates describing the situation; the superior then takes on the responsibility for resolution. If we imagine an identical copy of Figure 9-1 occupying the role of the supervisor, the entire discussion repeats only one level up. In many cases, the resolution is automatic, perhaps requiring negotiation with the peers of the C^2 nodes that issued the alert. In other cases, replanning may be required. In the most severe instances, yet another level in the hierarchy may become involved. Understanding the behavior of one node in the hierarchy allows the inference of appropriate behaviors in all the others. It is this replicate and customized feature that enables effective and flexible application of the proposed system design.

The Interaction Between C^2 Nodes

The interaction between C^2 nodes will become clearer through viewing the hierarchy in action, as in Figure 9-2, a notional diagram based on the OOTW scenario introduced above. The lowest level shown represents the individual mission packages formed to accomplish specific tasks (refueling, troop or victim transport, reconnaissance, and so on). Above that level are the wings (or other intermediate command entities). The CJTF sits on top. Significant events can flow either up (loss of aircraft, new target or mission opportunities, changes in weather) or down (commander's intent, policy change, shift of resources to other theatres) in the hierarchy.

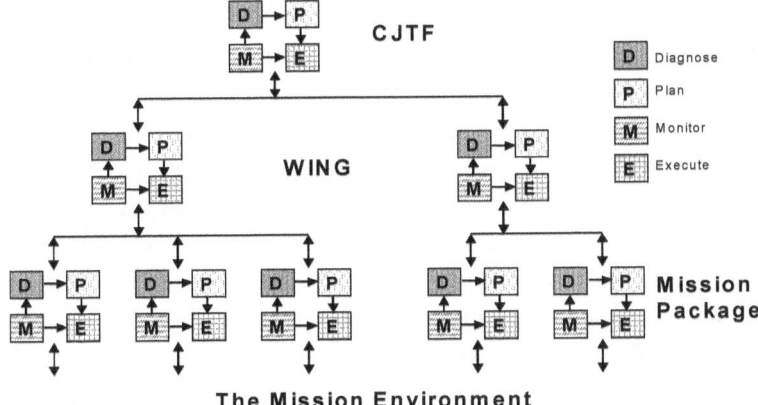

Figure 9-2. The HC2A hierarchy in action depicts a clearer interaction between C² nodes.

In the bottom up case, events can occur in the field requiring a response from the C² support system. Ordinarily, the status reports are normal, and the monitor function passes state data along for execution without changes. Sometimes, however, local replanning is necessary, for example, a change of course in response to local weather patterns, or substitution of one ambulance because another has a flat tire. Such changes can be made entirely locally, without the need to notify or request action from superiors. Sometimes, however, action at the wing level is required. For example, a key crewmember becomes ill and is unavailable for action. At that point, a request percolates up to the wing, then down to other mission packages looking for opportunities to reallocate resources. Here, an event detected by one mission package may end up affecting others, via the up-then-down path through the wing C² node.

If the circumstances have high-enough priority, or are sufficiently disruptive, it is easy to image this event's effects extending up yet another level, to the CJTF. In such an extreme case, and under the right circumstances, an event that occurs at the strike-package level could generate a replanning that affects several other wings. While such an incident is possible in theory, a good design will attempt to minimize both the likelihood of its occurrence as well as the breadth of its possible impact. Here is where notions of stability and plan inertia, both discussed in the scenario, come into play. A poor design will attempt global reoptimization even for small or incremental improvements in the objective function. A better design, however, will account for the disruptive effects of such replanning directly in the objective function, thereby keeping decision making as local and nondisruptive as possible.

With a top-down interaction, however, the CJTF can decide to change priorities or time lines, requiring lower levels in the HC2A hierarchy to respond appropriately. At any point in time, a subordinate C^2 node may receive a command from its superior to begin a global reoptimization process. The C^2 node will, in turn, pass this command down to its subordinate C^2 nodes, so that, in principle, all C^2 nodes in the entire HC2A system could be participating in a global reoptimization. Thus, the HC2A not only reacts to low-level events in the theatre of action, it also responds quickly and efficiently to top-level changes in guidance, strategy, or priorities.

In such a system, two key metrics of performance are the frequency with which low-level events percolate up in the hierarchy, and the elapsed time taken to complete a global replan. The HC2A prototype therefore includes software to monitor and report on such system performance metrics for subsequent off-line analysis. The possible further step is to analyze such data in real time, detect when the C^2 support system itself appears to be wandering away from good nominal performance, and then introduce the appropriate parametric changes (increase stability, change optimality thresholds). In this way, the HC2A system is subject to monitoring and control—a novel but powerful meta-controller for the C^2 process per se.

SKELETONS, MUSCLE, AND TISSUE

If the hierarchical structure is the skeleton of the system, then modeling and algorithms are its muscle and tissue.

We saw that each C^2 node considers those below it (or at the operational environment) as a system to be controlled. In that sense, the node needs predictive models of lower-level behavior, an appropriate set of commands available for issue (or, a command language), and a way of communicating objectives and relative priorities. Moving further down in the hierarchy, these models become increasingly more detailed; by contrast, upper levels will use suitable abstractions that capture only the features appropriate there.

In addition, as information is exchanged between levels, a process of consolidation and value-added takes place. It is not raw data that flows up, but rather data plus analysis—integration, fusion, causal connections, and so on—that have been recast in terms that the models and algorithms used by the superior can understand and process. Thus, HC2A functions not as a mere communications network, but instead as a processor, transformer of data, and producer of information. Negotiation of objectives, in effect, improves models used by the upper levels.

By analogy, consider pure data as colored red, while pure information (that is, conceptual content) is colored blue. The lowest levels ought to be colored mostly red, while the highest levels are colored mostly blue—with a suitable mixture of both at intervening positions. As concepts enter at the top, they are pure blue, turning increasingly red as they flow down toward the tactical components. Similarly, raw (completely red) data enters the system from below, and as it flows up, level-to-level, it acquires increasing conceptual content until, at the highest level, it is nearly all blue.

Considerable effort has been spent to develop models with

operational fidelity. A partial list of factors taken into account for air operations includes:

— Logistics (fuel at base, crew training and rest, weapons inventories, in-flight refueling, detailed mission time lines)
— Weaponeering (weapon/target effects matrices, weapons inventories, probability of destruction, level of desired destruction)
— Threats (SAM coverage and engagement models for a variety of SAM types, including fixed and mobile)
— Countermeasures (jammers)
— Routing (way points, cumulative threat, probability of survival calculations)
— Package composition
— Target valuation (nonlinear cumulative effects, time sensitive targets)

The HC2A structure accommodates detailed, validated models of military engagements—models on which a commander can rely [4]. Related to accommodating models, HC2A also encourages the development of an event-driven simulation of the evolving plan. While initially developed as a testbed on which algorithms could be tested and refined, this simulation capability is equally suitable as a predictor embedded in the monitor functionality. That is, the predictions of the simulation—with respect to accomplishment of objectives, risk, and use of resources—can be used to generate residuals for transfer to the diagnosis function. As in the scenario, poor models (if trusted without additional human monitoring and validation) can lead to mistakes or missed opportunities.

For algorithms, a key objective is scalability—the ability to handle problems of increasing size, with increasing levels of complexity, in suitably constrained time frames. The goal in HC2A is to divide and conquer: rather than solve a single huge problem at the top level, HC2A breaks the problem into smaller pieces (geographically, temporally, functionally), then solves the pieces in parallel (concurrently) lower levels in the hierarchy [5]. Breaking the problem into small pieces also enables the use of more detailed models maintained by subordinates.

The higher controller makes an initial guess at partitioning for both objectives and resources, then hands the guess off to the lower-level controllers. In turn, they return not only their proposed solution to the problem, but also, given a slightly different set of constraints or resources, sensitivity information (gradient estimates) about how they might improve their solution. Using this solution, the higher controller reformulates a partitioning—hopefully closer to the true optimum it seeks. Rather than accept a quick fix, the process repeats until there is no additional improvement in the quality of the solution. Conceptually, a continuing dialog takes place between the higher controller and its subordinates, with problem formulation information flowing down, and solution/sensitivity information flowing up.

As a practical matter, perfect is often the enemy of good enough. An initial rigorous Integer Programming (IP) solution proved to be computationally intractable, and was replaced by a set of well-designed heuristics to improve performance while maintaining a high level of modeling fidelity [6].

THE FACTORS TO CONSIDER

This technology contains a number of factors specifically designed to facilitate intuitive human interactions, and meshes well into a hierarchical military operational environment. While the scenario highlighted some of these factors, there is a more technical side as well.

Regarding the issue of target (or mission) value, as with other approaches discussed in this book, HC2A relies on optimizing an objective function. Given the range of command options at the technology's disposal (a large trade space), and subject to defined physical and operational constraints (for example, rules of engagement), the algorithm selects the command option that produces the highest value for the constructed function. Since this function is to be used as the sole criteria for selection of control, it must include:

— Probabilities, either as distributions or as expectations
— Values, or priorities, possibly time-dependent
— Assets, and the risk to them, including, perhaps, human life
— Strategic importance, of particular geographic regions, for example

Here, the objective function is the mathematical representation of everything that is important to the commander; there is no other way the algorithm has of deciding what's good and what isn't.

Beyond the mathematical difficulties in solving for the optimum plan, there are also practical problems in setting values for the parameters that enter into the objective function. How does one weight (assign a number to) the relative importance of different targets or missions? The proposed solution is assigning priority by type, geographic region, and phase in the overall campaign. That is, rather than requiring an operator to assign individual values to individual entities, it requires instead the relative importance of classes of targets—broken out by time phase and by geography.

Experimentation with such an approach has demonstrated that, when presented with information in this form, experienced military personnel are comfortable. Further, numerous simulations have shown that accomplishing objectives that are achieved using this scheme corresponds well to the intent of the operator who entered the data [7].

As part of the evaluation criteria used by the objective function, such a target-class approach will compare a proposed new plan against the current version and compute a measure of plan-to-plan variability or stability. To be good, a proposed solution doesn't merely have to use resources efficiently. It must also use them—to the maximum extent possible—in ways that it is already committed, that is, take into account the inertia factor. Further, the extent to which this inertia factor is employed is selectable—in some experimental simulations, it has been turned off completely so as to compare results with and without. In most cases, increases in stability can be had with little or no impact on performance, as represented by the value of the objective function.

ONE PLACE TO FIND A HOME

As with other technologies under discussion, HC2A makes certain assumptions about the infrastructure and military context in which it will be implemented and used. In many ways, HC2A is an extension of current C^2 practice. From the point of view of supporting infrastructure, it is evolutionary rather than revolutionary, a smooth step forward rather than a quantum leap. At the same time, the architectural core of this technology could also serve as the basis for some quite revolutionary approaches. For example, during recent research efforts, technologists were successfully able to interface to, and perform joint experiments with other researchers, itself a strong indicator of the approach's generality. Thus, HC2A could serve as the architectural framework within which many of this book's ideas could find a home.

Regarding communications, while a detailed engineering analysis has not been performed, all indications are that communications needed to support an HC2A implementation are well within the means of the current technology. Certainly, the amount and frequency of data exchanges to support the optimizing algorithm are modest, and the communications interface to the sensors and military intelligence system also appears quite practicable. For example, HC2A is adequate to support a four-hour strictly synchronous replanning cycle, for example, which requires no more military intelligence bandwidth than is currently available. As additional requirements and capabilities come online, the system should be able to scale well to accommodate them. In short, the current state of the art is more than adequate to support a very powerful version, one that can then scale nicely as technology continues along its expected exponential improvement rate.

As far as sensor support, military intelligence dependencies were not a central focus of the research. Thus, while hard data on this topic is not available, it would certainly be one of the first research areas should there be interest in pursuing this technology. Generally, the HC2A design philosophy accommodates current practice, while at the same time providing scalability as additional capability becomes available. One open question concerns how the hierarchical control and algorithmic architecture, shown in Figure 9-2, maps onto a corresponding architecture for intelligence gathering and processing. Given the divide between intelligence and operations, how does data entering at a given level flow down, or up, to other levels, and what intermediate processing or exploitation must be accomplished? How might HC2A make best use of—and exploit—a globally accessible common operational picture? While these are important and interesting questions regarding mid-and long-term future deployment, a very powerful HC2A implementation, one that relies on the current state of military intelligence timeliness and quality, appears to be feasible.

Considering HC2A's impact on the military organization, the technology's architecture has the flexibility both to mimic, and fit into, the traditional military hierarchical command structure. The C^2 nodes, which are abstractions in the HC2A schema, map nicely onto real military organizational entities. Officers, who already understand the existing structure's current functionality and products, will look to an HC2A implementation to take over known roles and responsibilities. However, it is also possible—even likely—that traditional ways of decomposing tasks and ensuring span of control and accountability will undergo changes in the coming years and decades. In that case, such a flexible structure as HC2A will provide additional benefits. Because of the modularity of a component like a generic C^2 node, it is hard to imagine a realistic command structure onto which HC2A could not easily map. While the HC2A approach does not, in and of itself, impose or require significant organizational changes, it is inherently flexible enough to accommodate them when and if such changes become necessary.

There is also the issue of unattended operations. One recent research goal in applying control theory to military C^2 problems is to see how far the idea of significantly reduced staffing requirements can reasonably be pushed. Increased autonomy provides additional potential benefits, in reductions in latency and cost, along with corresponding improvements in responsiveness and efficiency. There is, however, justifiable skepticism among the experienced officer corps about the practicality of such a vision, and the burden is rightly on the research and development community to make a compelling case.

On the issue of unattended operation, HC2A is cheerfully agnostic. While most research efforts to date have focused on unattended operation, it's simply a choice of research priorities and not a restriction imposed by the architecture. Even at this early point in the development process, all system actions are logged for subsequent off-line analysis, and supporting tools have been built. In addition, some online monitoring tools exist, demonstrating how most events are handled lower in the hierarchy without the need to percolate up. These tools operate synchronously with the underlying event-driven simulation—the experimental surrogate for a real battlespace—which in turn runs 100-500 times faster than elapsed time in the real world. Thus, monitoring system behavior is already well in hand.

As for actual human participation in the construction and selection of command decisions, while the architecture supports such interfaces, such participation has not been an emphasis of research to date. Indeed, the need for such a capability increases higher up the HC2A tree of nodes. Further, within a given C^2 node, it appears that the diagnosis and planning functions are the most likely places where such a human interface capability might be beneficial.

FOUR-WAY INTERSECTION

From a technical perspective, what remains to be done before HC2A can be brought to operational status? There are four areas of consideration: modeling, algorithms, open-loop versus closed-loop issues, and active adversaries.

Modeling

By current operational standards, the status of models employed in HC2A work is credible. It is clear that the architecture is able to accommodate and make use of detailed models. Given the current state of the art in model development, however, creation, initialization, and maintenance of these models is likely to remain a time-consuming and expensive process.

Algorithms

Concerning optimization, the algorithmic development provides a clear example of the dilemma facing much of the optimization work in C^2 support systems. An initial IP formulation, which exhibited considerable mathematical rigor, completeness, and formality, proved to be computationally infeasible. Faced with the choice of relaxing modeling fidelity (to improve run-time), or of introducing heuristics (to reduce the search space), the team wisely opted to preserve modeling fidelity. The result is a system that provides very good and practical answers, but can no longer make claims for rigorous optimality.

Regarding robustness, for realistic models of military C^2 problems and implementations of optimization, in the face of modeling errors, formal control theory cannot provide guarantees of robust performance. However, in the face of unmodeled discrepancies, the employment of state feedback in HC2A has been shown to be highly effective in maintaining robust performance [8].

Open-Loop or Closed-Loop?

Is the HC2A system open-loop or closed-loop? It seems closed-loop in its ability to respond (via replanning) to unforeseen or inherently probabilistic events. In applying an approach akin to MPC over a finite time horizon, accepting feedback about the state of the battlespace, and adjusting plans in response, HC2A clearly closes the loop.

However, HC2A has something else at stake. The question is whether the HC2A controller knows about itself. That is, in applying its optimizer, does it explicitly take into account the fact that it, itself, is present in the system—and will be able to respond to unforeseen events whose stochastic properties are known? Put simply, in formulating its plans, does HC2A model its own presence in the system?

For the work done to date, the answer is no. While the system does respond rapidly and asynchronously to triggering events that require replanning, the optimizing algorithm is open-loop. As the engagement proceeds, the algorithm relies heavily on the frequency and accuracy of state feedback to adjust its courses.

While this is indeed the current state of the optimizer, it is not an unalterable feature of the architecture. Indeed, joint experiments with the technology described in Chapter 7 demonstrate that, when the problem size and complexity are reduced sufficiently to make such a technique practicable, the HC2A approach is compatible with a fully closed-loop stochastic optimizer. Further, it remains an open question regarding how much additional practical advantage a closed-loop approach provides. From a practical perspective, just as suboptimal is quite often good enough, so also the advantages of a closed-loop MPC approach have yet to be rigorously justified in the real world.

Active Adversaries

Although modeling an active, intelligent adversary is another omission in the HC2A approach, it may not be a significant shortcoming in a number of situations, such as the OOTW scenario described earlier. In such scenarios, where opportunities for adversarial action and maneuvers may be limited, failing to model adversarial strategy and planning activities explicitly may be of little consequence. In other scenarios, the way in which human operators specify objectives will address adversarial responses implicitly.

Finally, can HC2A serve as a platform within which game-based adversarial models and optimizers can comfortably exist? The issue appears to be whether the adversarial models and optimizers can be decomposed along hierarchical lines, in the same way that resource allocation problem was decomposed. Should game-theoretic approaches be incompatible with a hierarchical architecture, this might severely limit the domains in which HC2A could effectively function. However, there is no strong reason to think such a decomposition is impossible; indeed, further research along these lines seems justified.

SUMMARY

HC2A is an architecture that achieves enterprise-wide optimization, while delegating decisions to the lowest level that can solve a problem, levels that have the best and most immediate information, and the most relevant expertise—which may not be available at a higher level.

HC2A architecture is composed of C^2 nodes organized in a hierarchy consistent with current military organizational practice. Each node has the same functions—monitoring, diagnosis, planning, and execution—but differs in details. Each node is a controller, and each one considers the rest of the hierarchy below it as the plant that it controls. If possible, a node attempts to resolve its control tasks without recourse to the higher level of hierarchy; otherwise it passes the information to the next-highest level.

By breaking problems into smaller pieces, the hierarchy permits the use of high-fidelity models throughout system, a use that would not be feasible if the entire problem were solved at a higher, global level. To arrive to an optimal decision, a higher-level controller makes an initial guess at partitioning objectives and resources, and then hands them to lower-level nodes. These nodes, in turn, compute and return their solutions, estimates of sensitivities, and suggestions on how repartitioning might improve the solution. Using only limited information about sensitivities, the higher node can make adjustments that optimize globally the overall solution.

HC2A's objective function penalizes solutions that result in large and frequent changes, leading to balance between stability and agility. Keeping decision making as local as possible also helps to minimize disruptions.

Key metrics of performance for such systems are the frequency with which low-level events percolate up the hierarchy and the response time to them. When performance deviates from desired, HC2A is subject to meta-control—adjustment of its parameters.

CHAPTER 10

A MARKET-BASED
TECHNIQUE IN C²

At 0800 hours, Operations Commander (OC) Sandra Smith sat facing her terminal screen with a fresh cup of coffee and the task of planning the 1,000-plus air missions to be flown in the next combat period. Scanning the flex-panel for the sorties summary list, she noticed that these included approximately 500 targets, 300 ground-force suppression missions, 100 refueling missions, and another 100 jamming support missions, all to be flown by a six-nation allied force. Yet OC Smith had limited knowledge of the other nations' capabilities. To make matters worse, a new force of highly classified intelligence collection (or black) assets had recently arrived, needing to be tasked, but Smith knew even less about those.

In such times of political sensitivities to ever-changing alliances, OC Smith couldn't afford to have the various allied forces, as well as her own units, send her sensitive details about what they had on hand or how they were planning their missions. Bottom line: for each mission, all she really wanted to know was could each unit do it, and what was its chance of success?

At 0815 hours, OC Smith stood up from her terminal, confident that a near-optimal allocation of tasks to units—an allocation that also incorporated an assessment of risk in the planned routes—had been reached. She smiled tightly: negotiations had been swift. In fact, they would have been faster, but risk assessments from the black assets had changed remarkably on the third round. By the sixth round, she had her solution: all 1,000 missions were planned.

How was that possible? First, based on actual risk information, her counterparts had been route-planning each mission on their local systems. That only made sense: after all, they knew more than she did about their assets' status and effectiveness. Second, they provided her across-the-board risk assessments (or bids). Then, to negotiate with her planning counterparts in the various allied forces, as well in her own units, OC Smith had run a software tool called Auctioneer. At each round of the auction, as Auctioneer provided the best allocation of tasks, each planner put in new bids based on assets and assumed risks. Finally, the planners and Auctioneer converged on a solution in just six rounds.

As she refilled her coffee cup, OC Smith recalled a story that her mentor used to share about needing hundreds of operators and a full 12-hour work shift to perform the same function that she completed in a mere 15 minutes. "We used to collect voluminous details about the situation and capabilities of subordinate and allied units," he had said. "But all those details were always outdated by the time we so laboriously compiled them. Our output was less optimal," he would shake his head, "and 30 to 50 percent of the tasking orders had errors in them."

ADVANTAGES OF THE MARKET

While the story above is a futuristic science fiction, the technology under development by O.R. Concepts Applied (ORCA), Whittier, California, may one day evolve into a high-tempo, flexible planning and replanning tool for operations commanders that addresses both route-planning and co-dependencies between missions.

The idea is inspired by the successes of computational and decision-making techniques patterned on economic market processes [1]. The technology, tentatively named Auctioneer, uses an auction to aid in information sharing and decision making, performing operational-level planning to allocate tasks to strike as well as support units from multiple locations to satisfy tasking requirements. Auctioneer sends task parameters (such as target location, desired level of destruction, and so on) to distributed tactical-level planners. For their part, the tactical-level planners perform route-planning and return probabilities of success and unit survival for each task received from Auctioneer. Auctioneer then uses all this information to optimize the task allocation and assignments to the various units.

The Auctioneer approach has two immediate advantages over conventional centralized planning processes. In today's Air Operations Center (AOC), route planning is done after the air tasking orders have been issued. Interactive autorouting and analysis tools (such as the ORCA Planning and Utility System, or OPUS [2]) are in use, but only locally, and by flight crews on a mission-by-mission basis. In an Auctioneer-based scheme, route planning is still done locally (for example, OPUS running on separate machines at each base), but it would be done at the same time that Auctioneer performs task allocation centrally. Moreover, since it is impossible to capture the intuition and experience of the operator performing the local mission planning, and since that operator also has access to information regarding maintenance and past performance of assets and crew that cannot be captured effectively and sent to a central location, each mission's route-planning and survivability calculations are still done locally. Finally, because Auctioneer allocates tasks based on the combined missions' projected success, the result is higher expected survivability as well as higher probability of campaign success.

Auctioneer's second advantage is a reduced requirement for information exchange. In future battlespace, it's likely that there will be a pressing need for fast, accurate, and reliable communication methods that will not require sensitive information exchanges, which in themselves will become increasingly difficult to protect. With Auctioneer, performing detailed mission planning at the unit level (the wing level in the air operations context) allows each unit to retain its own classified data on survivability. Indeed, the information flow during the negotiations is reduced to task descriptions in one direction and simple probabilities (numbers between 0 and 1) in the other. The result is a compromise between the need to minimize information for security purposes and the need to maximize information for optimal planning.

BRINGING DECISIONS CLOSER TO THE SOURCE

Auctioneer relies on a distributed decision-making process that, simply stated, brings decisions closer to where knowledge and information reside. With such a process requiring a distributed architecture, the Auctioneer's envisioned input/output is illustrated in Figure 10-1.

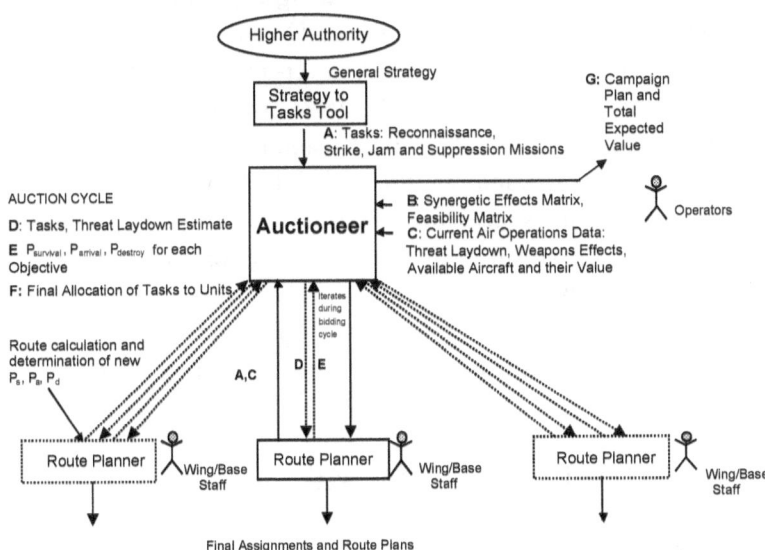

Figure 10-1. Auctioneer's data flow and distributed architecture bring decisions closer to the level where knowledge and information reside.

Inputs A through C illustrate the flow for the initial data required for each planning session. The tasks (input A) are reconnaissance, strike, and jam/suppression missions, which include the identity and location of the targets or enemy threats. The Synergetic Effects Matrix (input B) indicates reward numbers that capture the missions' relative values and the benefits or negative effects gained by combining any two missions. The Feasibility Matrix (also input B) is an array of feasible task/unit combinations (such as, 0 or 1 for each task/unit pair). Certain missions may be ruled out for certain units, either because they are too far away or do not have the required munitions. In addition, tasks can only be allocated to units if the units can feasibly accomplish the task. Current air operations data (input C) provides the type and location of known threats for the relevant geographic areas as well as the availability and relative worth of friendly assets.

The data flow for the iterations during a planning session is illustrated in Figure 10-1 by output D and input E. At each iteration, Auctioneer sends a problem set-up (output D) to each route planner, including threat laydown, weapons effectiveness, and mission information. The route-planning technologies return bids (input E), which indicate the probabilities of survival, arrival at the target, and successful mission accomplishment. After the auction mechanism has converged, Auctioneer produces an allocation of tasks for each unit involved (output F), and an expected outcome (output G), as a total of the expected rewards, which it sends with the complete allocation to the planning staff.

GOING ONCE, GOING TWICE . . .

Coupled with the distributed architecture, the Auctioneer employs a dynamic decision-making process. At the start of a new planning cycle, it is assumed that a higher authority, with the aid of a strategy-to-tasks tool (see Figure 10-1), supplies a list of tasks with reward values, if performed individually, and the relative worth of available friendly asset types [3]. Operators on the commander's staff enter Auctioneer's initial data, populating the Synergetic Effects Matrix

(SEM) and Feasibility Matrix (FM). While the former is an important mechanism that commander's staff manipulate to produce cooperative action in friendly units, the latter may eventually be populated by a software tool that checks aspects such as range to target and weapons effectiveness.

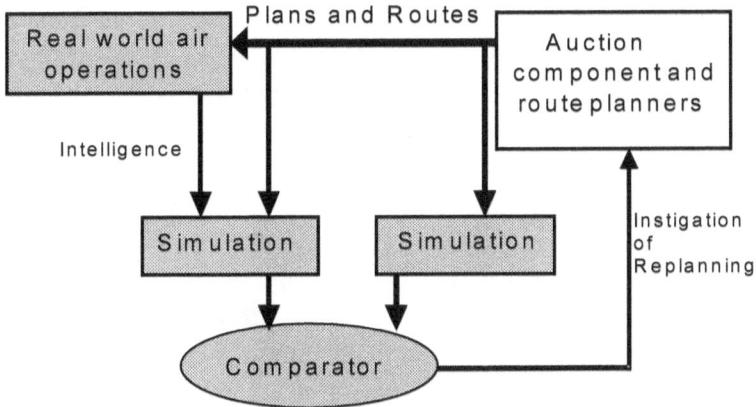

Figure 10-2. The bidding cycle between Auctioneer and the distributed route planners includes a series of iterations alternating between allocating for strike missions and for support missions.

The auction process then begins. The auction process consists of a series of iterations alternating between allocating for strike missions and for support missions as described in Figure 10-2. The Auctioneer operators begin the first strike mission planning stage by sending strike tasks, including target location and nature, estimated threat laydown, and weapons effectiveness tables, to the route-planning technologies distributed in the subordinate or allied units. Local operators (for example, officers distributed in the units) then use the route-planning technologies to calculate the probabilities of arrival (P_a), survival (P_s), and destruction (P_d) for each of the targets they can prosecute. The operators then return these values to Auctioneer as bids. Auctioneer then selects a tentative allocation that maximizes the expected value of the SEM rewards (based on the probabilities of mission success, P_a x P_d), while taking into account the insurance cost per aircraft (computed as $[1-P_s]$ times value of the asset) and the constraints in the Feasibility Matrix.

Auctioneer operators then move to the first support mission planning session, soliciting bids for these support missions. Similar to strike planning, local operators use the route-planning technologies and submit bids. Based on a similar expected value maximization to that of the strike planning—that is, taking into account the probabilities of success for the new support missions and the additional rewards in the SEM for support missions that aid strike missions—Auctioneer then selects an allocation for the support missions.

The process then iterates until it converges on a single allocation of tasks to units. Once again, the strike missions are replanned, this time based on the new tentative support mission allocations. Here, a new tentative allocation might be reached for two reasons. First, support mission allocations alter the estimated threat laydown; for example, a support mission to suppress a surface-to-air missile site makes routing over that location safer in the next planning round. As such, bids P_a, P_s, and P_d that are calculated by the route planners may differ from those in the first bidding session due to the new threat laydown. Second, the reward scheme for the strike missions may change due to the new set of allocated support missions; that is, strike missions that have synergetic value with tentatively allocated support missions will now be of higher value. Such differences, between one iteration and the next, diminish rapidly, converging after a few cycles.

As the iterative process between strike and support task allocation progresses, both Auctioneer operators and local operators learn more about projected missions and the team's capabilities. At any point before the full planning cycle is complete, the commander can make partial allocations, as needed for immediate action, or let the iterations converge to a fixed point—where no changes in allocations are produced by further bidding cycle iterations.

At each planning session, it is envisioned that Auctioneer will produce a plan of action for the next 12 to 36 hours. Replanning will occur either when a triggering event has occurred in the battlespace, or on a cyclic basis—expectedly, every three to four hours, but possibly as short as every hour. For every plan, Auctioneer produces results in 15 minutes or less.

The commander also manages Auctioneer's overall use. In a given air operations scenario, for example, the enemy C^2 center might be the primary objective. Because of heavy air defense, it is plausible that no route planner bids a strike mission to that target with a high probability of success. As such, the commander might change the valuations in the SEM, so that a first wave is used to find and destroy surface-to-air missile systems. Then, if those missions are successful, the commander might subsequently run Auctioneer through a new planning session, probably expecting that session to produce strike missions to the AOC that have a much higher probability of success.

When Auctioneer is used to replan for an ongoing operation, it assumes that while the replanning cycle takes places, the assets continue to follow previously issued tasks. The new allocations and route-plans created may then result in retasking assets.

In summary, Auctioneer technology leads to many attractive advantages, including decision making that has been distributed to take full advantage of local information and expertise, and an information flow that has been radically reduced. At the same time, despite efficient computation and seeming competition between units, solution quality and synergy between tasks is achieved.

SYNERGY IN ACTION

Even in a centralized planning process with human planners, it is challenging to construct tasks in a way that produces a synergetic effect; for instance, planning an air suppression mission that supports preparing a corridor for a strike mission. While the challenge of achieving such a synergy does not get any easier in Auctioneer's distributed and somewhat competitive process, Auctioneer's researchers have developed a mechanism that drives its process toward synergetic solutions.

Here, to encourage and reward cooperation among missions, a Synergetic Effects reward scheme is used, giving bonus values for synergistic missions. In the air suppression mission example, Auctioneer would give bonus values to the combination of an air suppression task and a strike task. In the auction algorithm, this idea is implemented through the SEM that lists additional rewards gained when one specified mission is combined with another. In the current simulation experiments, the SEM only captures benefits pair wise, which are implemented in generic terms: no value, some value, high value, extreme value, and critical value [4]. The SEM expands on the concept of a prioritized target list, providing structure to the decision-making process by allowing operators to consider interactions between missions, availability of resources, and possible allocations, all in addition to target valuation.

Further, it is envisioned that the main values entered in the SEM, which reflect the values of missions taken in isolation, would be provided by such higher authorities as the Joint Force Commander. Populating the matrix's other elements would then be a technical activity, performed by a commander's staff with a software tool. Clearly, not every entry in the matrix needs to be set in this process; instead, most entries will be set to a no-value default (for example, pairs of objectives that are independent and have no synergetic value).

Taking into account the expected value maximization that takes place in the auction, SEM values effectively direct Auctioneer's search—and so have a direct impact on solutions. Therefore, an operator can use the SEM values to capture heuristics, thereby guiding Auctioneer to an appropriate solution.

MAXIMIZED VALUES, SKEWED RESULTS?

Auctioneer performs a simultaneous determination of routes and task allocation but does not calculate an overall optimal solution. The simultaneous optimization of these would actually require a hybrid calculation of very high complexity: for every feasible combination of tasks to units, the calculation would have to render the optimal set of continuous routes for all airframes. Indeed, because such optimal routes are interdependent—if you change the flight path of one aircraft, so that it flies closer to a SAM site, such a change can affect the optimal route for other aircraft—each allocation requires an extremely high-dimensional optimization.

Auctioneer alleviates this calculation's computational burden by iterating between route-planning and task allocation, thereby efficiently producing a suboptimal solution. By computing new probabilities of arrival and destruction, which in turn alter optimal task allocation, each route-planning stage affects the subsequent task-allocation calculation. In effect, each task allocation stage reassigns a subset of the tasks, which in turn alters the threat laydown used in the next route-planning stage. Convergence to a single solution is guaranteed because of the auction mechanism employed [5].

Small-scale examples—single-digit numbers of aircraft and objectives—simulated in the current experiments have yielded reasonable solutions, at least on this scale. Throughout, the bidding scheme appears to achieve a pragmatic compromise between optimality and computational efficiency on the one hand, and stability of the iteration process on the other.

Still, because the current Auctioneer algorithm is based on an exhaustive search over possible allocations, it clearly has limitations. While the route-planning computation has been proven to be tractable—OPUS can produce these relatively quickly—when presented with large sets of tasks and units, Auctioneer's calculation becomes extensive and demanding. As research continues, it is envisioned that the expected value optimization routine will be redesigned to be much more efficient—and scalable to the order of hundreds of tasks and units.

There are other concerns as well. Maximizing expected value can lead to solutions that are skewed by high value/low probability missions; for example, a computed solution might indicate that all aircraft should attempt potential suicide missions for extreme value targets. While different numeric SEM target values for the generic terms no value through extreme value might offset such skewed solutions, there are also several other methods to arrive at more balanced solutions. First, because operators are always involved in the planning process, based on outside information (the cost of going after a low-probability, high-value target, for example), it is possible to realign the computation, by altering the SEM or the Feasibility Matrix. Other possibilities include adding nonlinear aircraft insurance costs, making target values a nonlinear function of P_a and P_d instead of a simple expected value, and so on.

In addition, just as Auctioneer currently considers the benefits of pair wise missions—by increasing the dimensionality of the data-structure (that is, giving three or more indices to the reward values), it could capture the benefits for combinations of three or more missions.

Similarly, by including timing information in the mission, the value of temporal sequencing of missions could also be captured in the SEM. For example, consider a reconnaissance mission with a final time of 1100 hours and a strike mission with a beginning time of 1100 hours. The corresponding value in the SEM pertaining to these two missions would then, in effect, capture the beneficial value of sending out a reconnaissance mission to the target and then following with a strike mission.

REPLANNING AND STABILITY

In the current simulations, it is assumed during the planning stage that all missions having allocated assets will be successful. This assumption is especially influential for future missions that involve corridors through enemy threat zones; here, route plans follow the corridor because it is assumed that threats in the corridor have been completely destroyed. Unfortunately, incorporating projected partial or probable destruction of threats into the route-planning process is problematic; as such planning greatly increases the complexity of computation of optimal routes. Again, the assumption of prior mission success for future planning is a pragmatic approach to mitigate complexity while allowing planning based on projected future outcomes.

A potential concern regarding the application of this technology is that replanning can be unstable, leading to dramatically different solutions from planning session to planning session. For unmanned aircraft, this so-called thrashing (for a more thorough description, see Chapter 3) may be considered a small point, but it would be highly unnerving for human aircrews and support. Therefore, in future implementations it may be possible to weight new solutions, giving an explicit figure of merit for keeping existing allocations unless there is a pressing need to change.

Auctioneer triggers replanning by two possibilities: the occurrence of a significant event in the battlespace or the build-up of a threshold delta in the threat laydown. Figure 10-3 illustrates an architecture in which a battlespace simulator and comparator are used to determine whether significant changes have occurred in the battlespace, including the threat laydown, since the previous planning session. Here, outputs from Auctioneer and route planners are fed to two simulations. In the simulation on the left, it is assumed that all plans go as expected; therefore, the nominal planned execution is simulated. In the simulation on the right, intelligence data, such as friendly attrition and success in accomplishing tasks, is also inputted to the battlespace simulation. The comparator block represents some decision-making process by which these simulations' outputs are compared to determine whether a replanning session is required. One preset standard, for example, could be when sufficient mismatches have been received between expected and actual mission outcomes to imply an incorrect threat laydown. Such a mismatch might result in either commander adding reconnaissance missions to the task list or route planners changing success probabilities. Such adaptability means that Auctioneer's models can consistently be updated based on intelligence and other data.

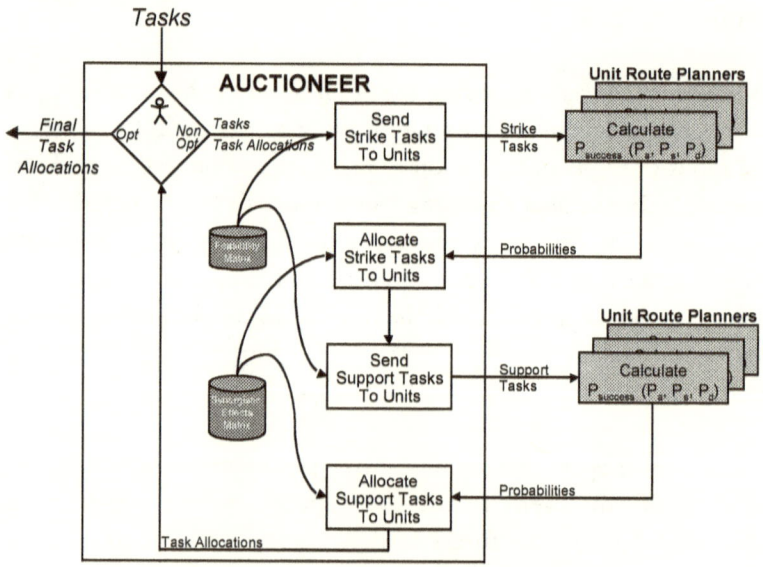

Figure 10-3. Expected task outcomes and actual task outcomes are compared to determine when to replan.

It is also envisioned that, based on such significant events as the appearance of time-critical targets, Auctioneer operators can invoke replanning sessions, as can higher authorities when objective lists change substantially.

WHO IS IN CHARGE?

Auctioneer's distributed, market-like nature also introduces specific challenges from the perspective of human factors.

Currently, considerable human involvement is required to decompose high-level campaign objectives into specific task lists. Here, a Strategy-to-Tasks tool could be developed to track the decomposition process, providing mapping from individual tasks to the objectives they support. Further, such a linkage would also provide some of the information needed to produce the SEM.

As high-level objectives change, the SEM, in turn, will require updating and modification. Such change can be normal, as an operation proceeds from one phase to another, or it can be in response to unexpected events. Currently, there are no automated mechanisms available to support such an analytical effort. Therefore, it appears that any updating or modification will be manpower-intensive, performed periodically throughout planning and executing an operation. Here, an automated tool, which captures subject matter expertise and applies predefined rules to SEM generation and maintenance, should be considered. As such, the tool must consider not only individual tasks and their mapping to higher-level objectives, but also how task groups can be linked to each other to accomplish stated objectives. Finally, the tool needs to help determine how defensive weapons systems affect risks associated with completing a task.

Even with such a tool, the process will rely on trained, experienced human decision-makers who will efficiently and effectively support the analytical process and make reasonable final determinations. Throughout, an organizational infrastructure must provide opportunities for individuals to receive the training needed to attain—and maintain—the skills necessary for these highly complex tasks.

Human involvement, requiring specific interfaces to facilitate inspection and possible modification, will also be needed to evaluate the auction process output, provide a final assessment of missions scheduled, and manage the auction process.

For its plan, Auctioneer provides an expected value, which establishes a baseline for measuring how well that plan is executed. In addition, a mechanism for comparing the plan's projected outcome to its achieved outcome, and for managing the plan's execution, provides a military decision-maker with the ability to assess performance at a relatively high level. Automated tools, based on the decomposition of high-level objectives into task/target sets, facilitate a drill-down capability to investigate performance against specific objectives. Here, tools should both provide the capability to add time-critical targets to the SEM and help an operator to evaluate Auctioneer's performance (in part, as a method for learning on the job and improving subsequent decisions).

In general, local interaction with route-planning technologies will be such that local users will not override either the route plan or any probabilities calculated by the route-planning technologies. In special cases, however, local users can override the plan or calculations, but based on sensitive information only available locally—for example, partial damage to stealth aircraft affecting radar cross-section, local shortage of specific ordinance, or local intelligence regarding the threat laydown. While such information can easily affect mission outcome, it might not be explicitly modeled in the route-planning technology, thereby requiring operator override.

A KNOWLEDGEABLE AUCTIONEER

Another crucial issue for Auctioneer's implementation is how it integrates with current and proposed information sources. Currently, it is envisioned that the intelligence community will provide geographic information, target database, threat information, and enemy weapons characteristics, while the operational community will provide a commander's guidance for objectives, priorities, and constraints (including, for example, the rules of engagement). Further, it is expected that the operational community will also provide friendly weapon characteristics in terms of effectiveness charts. Therefore, input for the SEM construction would necessarily be a collaborative effort between the operations and the intelligence communities.

Several databases need to be developed to support using Auctioneer. First, a target database would be required [6]. Second, a database that links specific targets to higher level objectives should be developed and maintained. In particular, while such a database would provide basic mapping of individual targets to objectives, it would have to be modified by time and spatial constraints, as reflected in operational objectives. As such, the intelligence community must quickly and accurately update individual targets' current status, particularly for damages, as they affect assigned target values.

SUMMARY

Computational methods inspired by market processes have been successful in solving complex optimization problems in variety of fields. Auctioneer pursues this promising direction as well, by using an auction to perform operational-level panning, allocating tasks to multiple distributed units.

This approach allows integrating task allocation and route planning into a simultaneous process, without requiring full centralization. Tactical planning can still be performed locally, bringing decisions closer to where immediate expertise and information reside. At the same time, information exchange requirements are drastically reduced—an advantage from the perspective of security and resistance to information warfare.

In a typical auction round, Auctioneer sends requests for bids—each task's parameters—to distributed tactical-level planners. These, in turn, perform route planning, then return their bids—each task's probabilities of success and survival—to Auctioneer, which then uses this information to optimize task allocation. The iterations alternate between allocating for strike missions and for support missions, and the process rapidly converges to a near-optimal solution. The same process is also used to replan and retask in ongoing operations.

To achieve synergy between tasks and missions—a difficult requirement—Auctioneer uses a mechanism called the Synergistic Effects Matrix, which, within the auction, rewards those solutions that combine cooperative missions.

Although the approach is not strictly optimal—the route planning and task allocation are not fully combined into one extremely large optimization problem—the bidding scheme is a practical compromise between rigorous optimality and computational feasibility.

CHAPTER 11

THE LANDSCAPE OF ROBUSTNESS AND VULNERABILITY

Consider the situation depicted in Figure 11-1. A rogue nation with a significant military capability, Outlawland, recently began flexing its muscles toward three neighboring countries: Eastland, Westland, and Southland. Then, under the guise of conducting military training exercises, Outlawland positioned large numbers of troops in Regions I-IV. Finally, two days ago, Outlawland forces crossed the national borders at all four regions, appearing to be approaching strategic targets in the three neighboring countries.

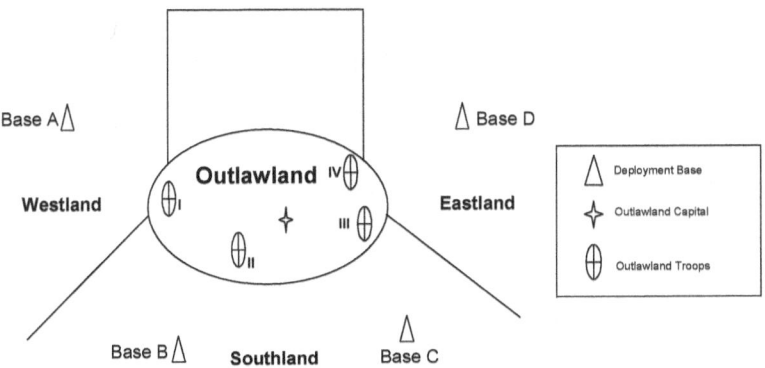

Figure 11-1. Regions of the Blue Coalition flank Outlawland Regions I-IV, where individual battles are fought by Blue assets from bases A through D.

The Blue Coalition (a mutual protection organization comprised of Westland, Southland, and Eastland), formed just six months ago in response to a significant increase in Outlawland's military capability and aggressiveness, assesses the rapidly worsening situation. Although U.S. military advisors to the Blue Coalition are already present in the theater, at this stage the U.S. remains an observer; it has not yet committed combat forces.

As the Coalition convenes, its decision-makers use emerging decision-support technologies that help determine possible options for initial and subsequent asset deployment, including specific options for bases A-D. In the present conflict, there are four distinct battles and corresponding battlespaces. An important concern in the Blue Coalition commander's mind is the effect of coupling individual battles in areas I-IV. Indeed, while individual battles in these regions may be concerned only with local objectives, coupling the battles poses additional challenges for the commander and forces—allocation of resources, movement, replacement, and so on.

Simultaneously, another key objective is to identify potential weak regions in the battlespace for both Red and Blue forces. It is therefore vitally important for the Coalition commander to survey the individual battlespaces, making assessments of Red and Blue vulnerabilities and strengths. At that point, the commander may take measures to modify Blue positions in order to exploit identified Red weaknesses. Here, the Commander seeks a *value landscape* of the battlespace: he wants to know how a change in battle parameters would likely affect Blue and Red forces. Even a high-level, coarse estimation of the value landscape will be extremely helpful.

At any time, Blue and Red units involved in a battle can be described by the values of their significant attributes, or *states*, and in the course of a battle various Blue and Red forces can naturally be in different states. Once a battle has begun, in order to steer the units towards desired goals, the appropriate controls are exercised for each Blue and Red unit's possible state. At any given time, the set of best controls for all units is referred to as a *strategy*. The collection of such strategies proves to be invaluable in obtaining a value landscape.

SURVEYING THE TERRITORY

To obtain a value landscape, a commander performs sensitivity analyses. Indeed, military decision makers continuously perform a wide variety of estimates: given the killing potential of certain weapons, for example, and the current combat situation, a commander makes estimates of the probability with which these weapons will achieve the kill. He makes estimates of the importance of his and the enemy's assets, of the costs of such operations as asset reallocation and task reassignments, and so on.

When a commander does make such estimations, he naturally needs to consider what may be the impact of errors or changes in his estimates. How will perturbations in battlespace parameters affect the battle outcomes? If, vis-à-vis estimation errors, there are large swings in outcomes—a steep gradient, or a spike in the value landscape—then a commander's forces could be vulnerable. If, however, the gradient is relatively flat, then the estimation errors are somewhat inconsequential—and a commander's strategies are robust. Currently, there are no readily available technologies, or tools based on such technologies, that let a commander answer such questions of robustness and vulnerabilities.

Sensitivity analysis, verifying strategy vulnerability or robustness, asks such questions as how vulnerable is a particular strategy to changes in kill probabilities? How robust is the strategy to changes in the importance attached to various assets? For the Commander, quantitative answers to such questions afford a high-level view of the situation, giving him a basis for confidence in his decisions during an operation. Of course, the results of such analyses may also indicate that a commander should change his strategy.

But how can a commander thoroughly review the possible impact of errors, or of changes in battlespace parameter estimates? In the Outlawland scenario, a strategy would be developed using a computer-based tool, then, using multiple simulation runs, sensitivity measures would be determined. Thus, for sensitivity analysis, developing strategies and running simulations would work together. As an example, Figure 11-2 illustrates the result of a preliminary analysis.

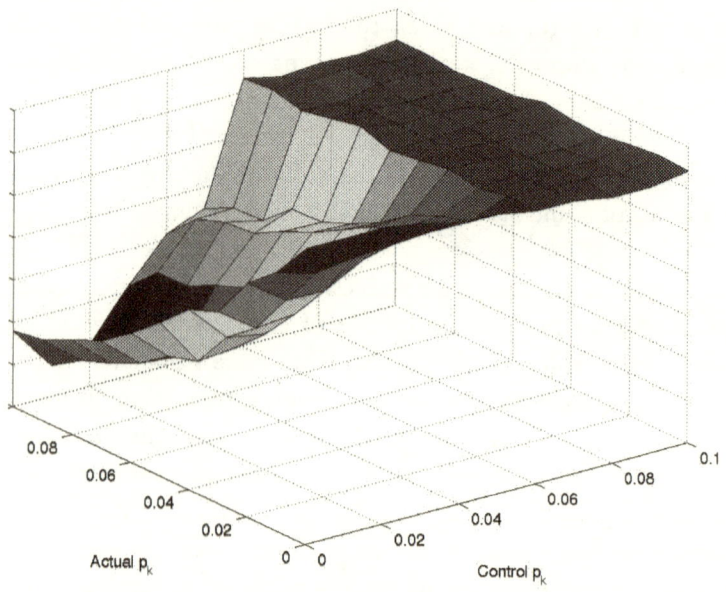

Figure 11-2. The surface depicts the number of Blue units that survived in the simulation of battle in Region II, as a function of actual and assumed lethality of enemy weapons.

In Figure 11-2, P_k indicates the Coalition's high estimates for enemy weapon lethality. If the Coalition uses appropriate tactics for a high-risk environment, Blue could maximize survival, as indicated by the plateau on the graph's upper right. If, however, the Coalition underestimates enemy weapon lethality, and correspondingly uses defenses or tactics inappropriate for a high-risk environment, Blue would have a much lower survivability, as indicated by the dip on the lower left.

Therefore, in this analysis, as long as the Coalition uses 0.06 as the minimum probability of lethality, it doesn't really matter what the actual probability is, for Blue would still maximize survival of Coalition assets. Here, the rapid transition region below 0.06 indicates a vulnerability region; whereas the plateau region represents robustness, or invulnerability to actual P_k variations. Based on such an analysis, then, the Coalition would be well advised to dedicate additional efforts to ascertain kill probabilities (the lethality of the enemy's weapons), specifically those on a transition point.

With multiple analyses performed over different parameters of interest, a commander would obtain a broad view of the value landscape for the operation and his intended strategy. He would gain insight into the sensitivity of the outcome as it relates to variations in assumptions and estimates. He would also identify those areas within his decision space that are relatively robust with respect to minor errors or mishaps, and those where even a minor variation may spell a dramatic change in events and outcomes.

REALLOCATION OF ASSETS

Another important issue facing a decision-maker in situations such as the Outlawland scenario is optimal allocation and reallocation of assets. A commander not only has to allocate his assets among the four bases, but he also must guess how the enemy forces may reallocate their assets. In the present situation, Blue assets are stationed at bases A-D, but can be reassigned to participate in operations overseen by other bases—decisions that will have to be made by the Coalition commander. How should the Commander allocate—and dynamically reallocate—assets between sub-regions? This set of problems, known as hierarchical allocation, has an additional component: in the midst of battle, how often should reallocation take place?

Clearly, the question of optimal controls for units, and the reallocation of units from one region to another, or one sub-region to another, should be governed by appropriate, well-defined objectives. However, the objective must be suitably formulated—formally, clearly, and unambiguously.

GAME-THEORETIC DECISION AIDS

North Carolina State University, Raleigh, North Carolina, along with sub-contractors Brown University, University of Georgia, and Tempest Technologies, have been developing technologies applicable to scenarios like the one described above [1]. Addressing the stated challenges, these technologies build the value landscape, identify both robust and vulnerable areas in the decision space, and provide help in initial allocation and dynamic reallocation of assets.

The decision aids developed by the researchers use robust optimization methods and rely on game-theoretic mathematical description and analysis [2]. Here, each force attempts to optimize its own strategy while recognizing that the opponent does the same. In such an approach, an objective or a goal is defined mathematically; while one side tries to maximize (or minimize) the value of that objective, the other tries to minimize (or maximize) it.

The power of game-theoretic techniques is that Red and Blue both attempt to optimize their strategies. Here, as each side tries to change the objective's value to its advantage, Red is not treated as a passive disturbance in Blue's optimization process, but rather as an active, intelligent adversary. Indeed, compared to treating an adversary as a random act or using doctrinal rules or scripts that Red should follow, a game-theoretic method allows for far more realistically modeling an intelligent adversary.

To that end, one technology, Action Advisor with a Sensitivity Tool (AcSenTool), analyzes scenarios with relatively few entities or groups of assets. Based on the degree of confidence in battlefield intelligence, AcSenTool suggests offensive strategies with potential alternatives.

Another technology, the Hierarchical Team Allocation Manager (HiTeAM), provides a means to decide how best to allocate resources across regions, and within sub-regions, for both Blue and Red forces. It adaptively optimizes asset distribution among the smaller battles that are coupled together. AcSenTool and HiTeAM have been integrated into what is called the Big Game—a prototype tool for decision-aid in C^2 of multiple small battles set in a larger operation. Thus, Big Game models situations like the Outlawland scenario.

Focusing on a few salient features of Blue and Red assets and actions, both AcSenTool and HiTeAM provide gross estimations of vulnerabilities and strengths. Not intended to operate on highly detailed, high fidelity models, the technologies afford a commander a coarse-level value landscape. Using detailed, high fidelity models, the broad analysis required to obtain the value landscape becomes computationally difficult.

Unlike alternative, heuristic-based methods, or pure simulation-based approaches, AcSenTool and HiTeAM use precise algorithmic techniques guaranteed to converge to solutions—if one exists at all—in the closed form. This means, rather than using such means as rules or empirical estimates that often lack clear definition of the circumstances under which they are valid, AcSenTool and HiTeAM employ precise mathematical manipulations to generate the best solutions.

ACTION AND SENSITIVITY

AcSenTool includes a core module, the Action Advisor, which recommends control actions (commands issued to the simulated units), such as attack SAMs, hide from SAMs, return to base, and so on for both Blue and Red forces. In addition, a simulation module performs sensitivity studies on Action Advisor's recommended control actions, issuing the commands in the course of a simulated battle. As a whole, AcSenTool helps make qualitative judgments about tactical actions in battles with small number of combatants. Although the current prototype of AcSenTool is focused on such assets as UVs, undertaking air strike missions, its concepts and algorithms can be extended to other types of assets and missions.

AcSenTool's operation can be illustrated using the Outlawland scenario. Here, a number of Red targets are in Region II, with SAMs to protect them, and Blue forces attacking these targets will inevitably encounter the SAMs en route. As such, the Blue forces, composed of groups of aircraft (or, units) originating from Blue base B, must decide whether to destroy the SAMs or merely fly over them on the way to the targets. In this case, Blue knows that both the Red SAMs and targets are immobile.

AcSenTool recognizes two discrete descriptors (or, states) for each Blue and Red unit: location/engagement and health/strength. For the Blue unit, the location/engagement descriptor is that moment's SAM encounter, while the health/strength descriptor could take any of four values—OK, damaged, severely damaged, or destroyed. For the Red SAM unit, the location/engagement descriptor is the particular Blue unit it faces, while the health/strength descriptor could take any of three values—OK, damaged, or destroyed. In the current limited implementation, the Red targets have no descriptors.

Thus, for example, at a given moment a Blue unit consisting of 5 UVs is in the state {7, OK}, meaning that it is engaged with SAM #7, and has not experienced much attrition or expenditure of weapons. At the next moment, the same unit might be in state {7, damaged}, meaning that it is still engaged with SAM #7, but has experienced significant attrition or is low on weapons. Perhaps the next state might be {11, OK}. And so on.

Modeling Dynamics

To describe the model's dynamics, state transition probabilities need to be specified. For each unit, a matrix must be created to specify the probabilities with which the unit will go from one state to another. Such probabilities depend on a number of factors, including:

— Number of Blue units attacking a Red defensive entity (e.g., SAM)
— Whether the SAM radar is on or off
— Whether the Blue unit knows that Red defensive entities are present
— Whether the SAM is damaged or destroyed
— Number of Red defensive entities flown over en route to an attack
— Unit maintenance status

At discrete instants of time, and according to probabilities in the state transition matrices, the algorithm determines the unit's change in state. Formally, such models are called Discrete Time Controlled Markov Chains [3], and the benefits of modeling dynamics with them are the model's simplicity (only current unit states matter in system evolution) and the availability of powerful formal methods to analyze such dynamic processes [4].

AcSenTool also requires specification of the battle environment. In this implementation, the environment is modeled via a *geography distillation*; that is, a simplified description of the relative geographic positioning of Red SAMs, Red emitters, and Red targets. In essence, if a Blue unit is located within SAM #7's sphere of influence, the algorithm thinks of it simply as location #7—without adding more detailed specification of the position. Such a description dramatically simplifies both the model and the number of calculations required for analysis. A route that a Blue unit pursues while it advances toward its target is similarly defined as a sequence of such distilled locations. For example, a route might be {2,7,11}, meaning that the Blue unit will proceed through the spheres of influence of SAM #2, then #7, then #11. Here, AcSenTool does not attempt to determine such routes, but rather obtains them from a separate routing tool. Outlawland Region II's geographic distillation is shown in Figure 11-3.

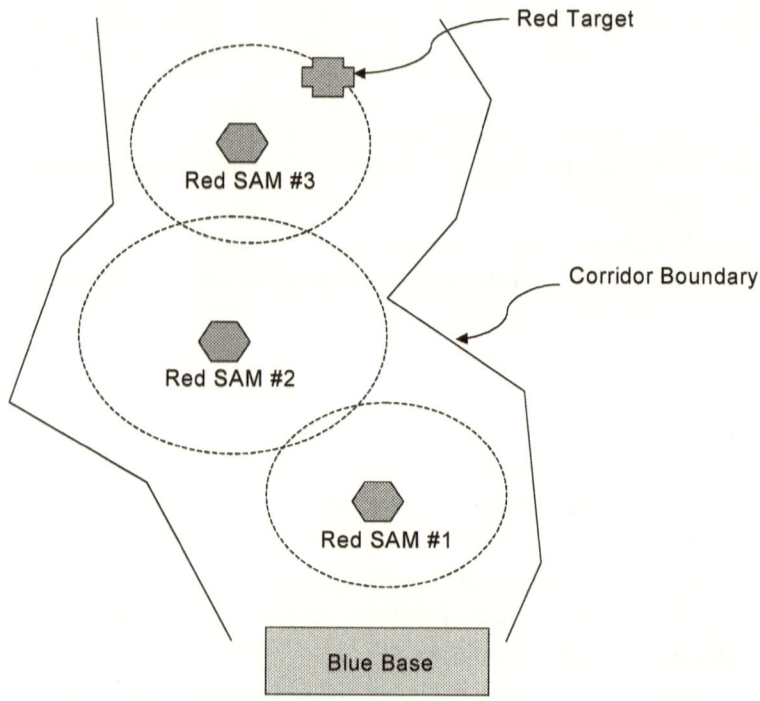

Figure 11-3. A geography distillation, showing the arrangement of SAMs with their respective spheres of influence, is required for AcSenTool.

A distillation is generated a priori and used by AcSenTool to determine the controls that need to be exercised over Blue and

Red units. In the current, simplified AcSenTool implementation, the Blue unit controls are to attack the SAMs, hide from the SAMs, or return to base. For their part, the Red SAM controls are to turn on or off the radar. By using a control algorithm based on game theory and backward dynamic programming, AcSenTool's Action Advisor component specifies the best control actions for all units. Such techniques are guaranteed to find unambiguous control actions for all possible states of the units. The control actions for all the Blue and Red assets are stored in control arrays that are subsequently used in the simulation.

Dynamic Programming

The control algorithm used by AcSenTool's core module, employs a dynamic programming (DP) approach to solve a game that has a mathematically specified objective involving Blue and Red forces [5]. Here, to solve the game means to determine the best actions that Blue and Red units would take—when Blue forces try to minimize (or maximize) the objective, and Red forces simultaneously try to maximize (or minimize) the same objective. As employed in AcSenTool, the DP algorithm takes the following steps:

1. The application of Backward DP principle involves starting at the final, or exit, time, with the desired outcome stated mathematically as an objective function (the value of which is called the exit payoff.) As one example, the outcome could be that at least three targets will be destroyed, at least six SAMs will be damaged or destroyed, and no more than three Blue units will be lost.

2. The algorithm works backward in time to find the strategies that lead to the desired outcome. The algorithm computes the *value function* at each time step; the resulting number tells the algorithm how far it is from the exit payoff [6]. In AcSenTool, this value function is defined as the average of

the payoffs for all possible state transitions from the current state of all units.

3. Since DP works its way backward through time, the iterations stop when the payoff obtained from the value function is same as the exit payoff. At that point, the algorithm is said to have converged to a solution. Then, the control actions for all Blue and Red units are the optimal control actions for all combinations of states of all units—an optimal strategy that is stored in the control arrays.

Monte Carlo Simulation

The Monte Carlo simulator, the accompanying module in AcSenTool, helps perform the sensitivity analyses. In the simulation, game evolution proceeds forward in time, and as the states of the units change non-deterministically, control actions from the controls arrays (created by Action Advisor) are exerted on the units. Multiple runs are performed, and statistics are gathered. Typically, as some parameters are perturbed (kill probabilities, for example), variations in outcome (average Blue unit attrition, for example) are recorded to make inferences about sensitivities.

Using the control actions stored in the control arrays and by performing multiple simulation runs (varying parameters such as kill probabilities each time), AcSenTool provides a means to assess the robustness or vulnerabilities of both Blue and Red forces. For example, a typical set of simulations may examine the remaining Blue forces after a round of combat—a number that may change as the kill probability is varied. If perturbations around a given value of the kill probability do not affect the outcome—that is, if the gradient is flat—then the Blue forces have some robustness in their strategy. However, large swings in outcome indicate a vulnerability or sensitivity to this parameter.

In the Outlawland scenario, AcSenTool would run multiple simulations to assess probable outcomes in Region II—as well as the other three regions. If estimates of Outlawland air defense lethality are uncertain, AcSenTool can analyze a range of lethality numbers to help determine the right strategy to pursue in each region. The results of AcSenTool control recommendations as well as corresponding simulations, the emerging value landscape would be similar to Figure 11-2. At that point, a commander would select a strategy that has a significant degree of robustness; that is, a strategy in which outcome is relatively insensitive to the lethality parameters. Alternatively, a commander could consider additional operational measures to increase Blue force robustness as well as to exploit Outlawland force vulnerabilities.

HIERARCHICAL ALLOCATION

The Hierarchical Team Allocation Manager (HiTeAM), a second technology discussed in this chapter, also uses game theory, but arrives at a different decision: the best way to allocate assets in a battle. In the Outlawland scenario, for example, there are four battles that the Coalition commander must fight in four regions— so that judicious use of assets is critical. The problem of allocating assets between such regions and sub-regions is called hierarchical asset allocation [7].

A tool based on this technology can be used to decide on an initial allocation of assets, or for determining the dynamic reallocation of assets during battle. Figure 11-4 indicates a possible battle situation in which Blue units would start from Blue base to engage Red units (here, SAMs and targets) in two regions partitioned into two sub-regions. Both before the battle, and as the battle progresses, HiTeAM would recommend how many Blue units should go to each region and to each sub-region.

In addition, HiTeAM can also specify a similar distribution for Red assets [8]. While in this scenario the Red targets are fixed, so there is no allocation problem for them, the SAMs can be moved among regions and sub-regions. HiTeAM affords the Blue commander a means to understand possible Red unit movements and deployments, and to know or guess the costs involved in performing such transfers.

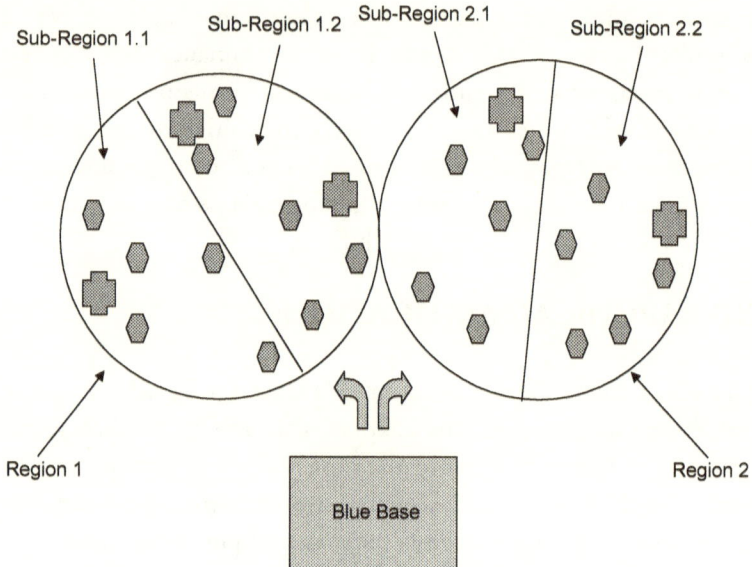

Figure 11-4. HiTeAM determines a hierarchical asset allocation among regions and sub-regions, while Red SAMs are moved across the region and Blue assets are reassigned to engage in the different regions.

At any time instant, these descriptors define the state of the overall system:

— Number of Blue units in each region
— Number of Red SAMs in each region
— State of the targets in each region

Exercising controls over the units means transferring some units to other regions. Of course, any transfer has the constraint that the total number of surviving Blue and Red units remains unchanged.

HiTeAM's objective function combines the requirement to optimize chances for success in the overall battle while at the same time avoids excessive costs associated with making necessary transfers of assets from one region to another.

In the Outlawland scenario, HiTeAM would help determine the initial distribution of air assets for strike and interdiction missions. Here, the main inputs needed are the number of Blue and Red assets, the costs of reassigning Blue assets, and the projected costs of moving Red SAMs to protect sensitive targets. The tool then recommends how many units should be assigned to bases A-D, and also makes predictions about how the enemy will redistribute his SAMs. Here, although HiTeAM does not delve into specific details of asset movements, it does assist in the overall decision making.

PLAYING THE BIG GAME

AcSenTool and HiTeAM may be used together in what researchers refer to as a Big Game. Here, the Monte Carlo simulator reacts to certain pre-specified events deemed important by the commander, such as reaching a certain level of Blue attrition. When an important event occurs in the simulation, the simulator reacts by triggering execution of one of the two tools, HiTeAM (to reallocate assets) or AcSenTool (to generate new, improved controls).

The Big Game corresponding to the Outlawland scenario has four concurrent executions of AcSenTool for the four regions I-IV. Prior to running the Big Game, the following information is required for each region:

— Geography Distillation—capturing the number of Blue and Red units, as well as the relative geographic partitioning of the Red units
— Transition Probabilities—identifying the probabilities with which the units will transition from one state to the next
— Parameters—deciding on weights (or, value) associated with

the various units, then depicting their relative importance in the objective function from either Blue's or Red's perspective; also included in this set are kill probabilities.

The purpose of running the four concurrent executions of the AcSenTool's Action Advisor is to get the best controls for all four regions. Then, the transition probabilities or the weights in the objective function are varied with each run of AcSenTool's Simulator to obtain sensitivity information. Based on that sensitivity information, a commander detects vulnerabilities in the Blue and Red forces and allocates resources accordingly.

A key benefit of the Big Game is that a commander is able to determine when it is appropriate to reallocate assets. For instance, the following key triggering events may occur during the battle's evolution:

— E1 = x number of Blue units remaining in region IV
— E2 = x number of Red SAM units remaining in region II
— E3 = x number of Blue units back to the base in region III

Having defined these events, it is then specified for each whether HiTeAM or AcSenTool is to be rerun and in which order. With these definitions in place, then, the Big Game is executed. As the simulation progresses, different events are triggered, producing a high-level view of when to reallocate resources and in what manner. Sensitivity tests are also run (by varying such parameters as transition and kill probabilities), obtaining updated vulnerability and robustness measures for all four regions surrounding Outlawland. Unlike the previous analysis for individual regions on a stand-alone basis, the current analysis shows the influence of coupling between the battles in different regions—the effects of possible reallocations of both Blue and Red assets.

CRITICAL BARRIERS

The new aspect of the present technology is the use of game theory to develop strategies—the computation of control actions using backward dynamic programming (DP) with a discrete, stochastic model. The fact that an objective function or goal of a game is specified up front, and that there is a DP approach to compute the control actions (guaranteed convergence to solutions if they exist), is noteworthy. Here an algorithmic technique will generate control actions unambiguously for all possible permutations of states of the participating entities without a reliance on heuristic rules.

The sensitivity analyses enabled by such tools may be utilized in a number of ways. For example, the value landscape information provided by AcSenTool, based on extensive Monte Carlo simulations, present a probabilistic unfolding of a scenario—and hence provide information about averages, average attrition of Blue assets, for example. Information may additionally be obtained on standard deviations as well. Under the right circumstance, such information could be of even greater military utility than average numbers. Increasing the uncertainty is a useful military strategy for one side or the other—especially if one is losing.

Admittedly, the current implementation of the two technologies described here is quite simplistic. Indeed, current research prototypes are limited in many respects, particularly in the types of entities described in the model, in actions, controls, and so on. These limits are not critical, for such models could be extended to a greater degree of fidelity. However, the fundamental features of the approach do call for making some difficult choices—and for accepting some important limitations.

Numbers Matter

In particular, the choice of states describing the units is critical. While the states should be sufficiently descriptive, so that the value landscape predictions are meaningful, there should not be so great a number of state variables that would naturally lead to a large computational burden. Low cardinality of the models—a relatively small number of entities, action types, and so on—is an important requirement for these tools to be effective. Here, difficulties arise with computational complexity, as well as storage and retrieval issues, when the number of entities and their corresponding states becomes too large.

Thus, in order to use this technology, only a few entities may be used, with a few, but important, corresponding states. This is where the fidelity question becomes paramount: will a scenario like Outlawland be viable using only a few entities and a few of their states? One instinctive reaction might be negative; however, if aggregated notions of states both sufficiently descriptive and numerically few are constructed, then such technologies as AcSenTool and HiTeAM might be both computationally feasible and useful. In view of the enormous complexity involved in taking all relevant factors into consideration, using such limited models seems only natural. Indeed, decision-support tools often rely on aggregated notions, like those presented in AcSenTool and HiTeAM.

Demands for Information

Other concerns regarding the practical implementation of such technologies center around the significant information requirements. One key element not well defined currently, for example, is the relative importance of individual targets in the area of interest. Closely linked is the concept of threat evaluation, which also changes over time. Given the dynamic nature of both elements, in a Big Game setting a very sophisticated simulation environment would be necessary in order to interact with such rapidly changing components. Another key issue is that commanders and staffs will need to be educated about—and trained to create—the correct formulation of objectives that are translated into mathematical expressions in the tools.

Further, there are a number of parameters and data, required for both Blue and Red units, read in as part of initialization:

— State transition probabilities of the Blue and Red units. As there is a significant difference in the characteristics of individual units and platforms, the determination of state transition probabilities is naturally a complex process. Here, a separate simulation system would most likely be required to generate transition probabilities on an individual basis for the units and platforms.

— Geography distillation is based on order of battle information available for either a real world or simulated engagement. The order of battle for enemy and neutral units, generated by the intelligence community, is usually a fused report based on various types of intelligence data. At an abstract level, this report provides a basic characterization of the battlespace. Because terrain features influence the spheres of influence of individual weapon systems, terrain features also need to be incorporated.

— Parameters. A consistent methodology is required to reflect the importance of assets, including weighting parameters— indicating the importance of Blue and Red resources—and kill probabilities. While the latter are doubtless affected by several factors, a high-level estimation of such probabilities might be a function of the concentration of Blue or Red forces in a region. Thus, for example, if many Red SAMs surround a Red target, then a Blue aircraft's kill probability is going to be small—a gross estimation, to be sure, but enough to provide high-level sensitivities.

Storage and retrieval issues are also important. The Action Advisor module of AcSenTool alone creates a control strategy file that specifies the controls that need to be executed for every state of the Blue and Red units. During simulation runs, then, for every state change of Blue and Red units the corresponding strategy (or, the controls to exert) are read in. Therefore, storage of large control arrays is an important infrastructure issue. If these control arrays are stored in databases, then quick and accurate methods of look-up and retrieval will be required.

HUMANS ON THE TEAM

Humans may have difficulty assigning quantitative values with regard to the importance of enemy targets. Further, they may encounter problems translating the commander's high-level objectives into quantitative values needed by AcSenTool's or HiTeAM's objective functions. Here, determining weights and coefficients in the objective functions could be challenging. Similar concerns apply to valuation of targets and assets.

It will also be difficult for humans to set parameters for sensitivity analysis—and to determine productive ranges for parameters themselves. To work effectively with the modeling and planning technologies, specialized roles and training may be required.

Finally, humans will have to judge the acceptability of such simulation results: if they are not acceptable, humans will have to decide what to change. To interpret the results of simulation, humans will require insight into the underlying assumptions in the model and the complex algorithmic processes that generate such results. Almost certainly, good visualization techniques will be required to understand the results of sensitivity analyses—and to detect vulnerabilities that may be hidden or inherent in the strategies.

SUMMARY

A value landscape depicts how the outcome of a battle depends on such important parameters as force composition and disposition, weapon capabilities, and so on. Steep gradients may imply great sensitivity to errors or unexpected events—that is, an area of potential vulnerability—while a flat landscape suggests robustness. Value landscape is produced by sensitivity analyses, and can provide a basis for commander decisions.

Within the multiple battle simulations required for the analyses, Red and Blue units must be intelligently controlled, which is where automated tools such as those described in this chapter are useful.

For simulated units, HiTeAM makes decisions about optimal allocations and dynamic reallocations of assets between battle zones. Using a game-theoretic formulation—meaning that the intelligent adversary's decisions are explicitly considered—HiTeAM finds the best actions available to Red and Blue at every given moment. Thus, HiTeAM determines the best possible reallocations and movements of Red assets as well as those of Blue.

As the simulated battle progresses, AcSenTool, also in game-theoretic fashion, commands optimal actions to the units. Using relatively simple models called Discrete Time Controlled Markov Chains, which offer powerful formal methods for analysis, AcSenTool employs a rigorous dynamic programming approach that solves problems for specified objective functions.

Together, AcSenTool and HiTeAM control Red and Blue forces in multiple simulations. The statistics collected in the simulations are used to construct the value landscape—an aid to a commander's decisions.

Although currently of low fidelity, such models can be potentially extended to a desired degree of realism. Still, the methods are suitable only for a relatively small number of entities, and should be used where aggregation can be applied.

PART V
TACTICAL COMMAND AND CONTROL

CHAPTER 12

OPTIMAL CONTROL AS A ZERO-SUM GAME

Six friendly-force aerial UVs, positioned west of the peninsula's center, have dual missions: to suppress enemy air defenses and to protect friendly manned aircraft. The UVs, orbiting in fuel-conserving mode, wait for the rest of the strike package, consisting of six F-15E (strikers) and four F-15C (escorts) aircraft. Divided into three sets of two, the lead UV in each pair is designated as the master of that pair and the master for the first pair is designated as the prime master, directing the master UVs for the other two pairs.

Each UV pair flies in a lead-trail formation with the trailing UV maneuvering in a cone between 4,000 and 6,000 feet behind the leader and 40 degrees to either side of its tail. The three pairs of unmanned combat aircraft, positioned in orbits, are separated vertically by 5,000-foot intervals. Using Global Positioning Systems, the UVs maintain their flight profiles, designed for maximum range, loiter time, and effectiveness. Although the UVs share their sensor information, when they operate in stealth mode all active emitters are shut down.

Previously, the six UVs were launched under cover of darkness from their deployed base. Preliminary instructions were downloaded into their memories, including their flight route, target hierarchy, ordnance and armament reserves, and current updated terrain models. In the air, the UVs' onboard sensors estimate the locations of enemy targets and assets. Then, through pre-established data links among all the vehicles, the UVs share the information with each other. In addition, sensor data is fed back to a mission-guidance controller at the Tactical Command Element (TCE).

This scenario, of a hypothetical future engagement involving unmanned combat units, indicates that once the UVs' memories were loaded with a program that commanded them to fly their courses, there was little control that could be exercised on them. A key difficulty is that the data link to TCE is not always reliable. In addition, manually controlling an UV—by issuing commands over radio links from a distant control center—is hardly effective.

EMPOWER THE TEAM

STRATEGIST (for Strategic Duelist), a technology that aims to address the scenario's problems, has been developed by researchers at Washington University [1], St. Louis, Missouri.

Offering a way to exercise control in real time, STRATEGIST works not from afar but from *within* the group. The role of a human controller at TCE would be to provide only general, high-level guidelines. All or most of the local controls would be relegated to a master UV in a group of UVs.

With such an approach, emphasis shifts—from ensuring absolute security and reliability of the data-links connecting the UVs with TCE—to developing smart-control algorithms on board the UVs. As the coordinates and other parameters of the various units change, a designated master UV performs an online, real-time computation of the controls. (Here, unit is used to denote an individual platform or a formation of platforms controlled as one group.) Such computed controls are sent over relatively short distances to other UVs in the team, providing real-time control.

STRATEGIST's developers have proposed a game-theoretic algorithm that would let a controller make real-time adjustments to the motion of UVs. Such adjustments, made by issuing commands to the vehicles, cause them to take certain actions (change course, for example). Further, such controls need not be limited to movements; they can also command the UVs when and how much ordnance to use against a particular enemy target, and to perform other actions as well. Based on internal mathematical models, and using sensor observations, STRATEGIST both provides online continuous tactics development and issues control signals.

In STRATEGIST, the game-theoretic algorithm specifies not only the best controls for the friendly UVs, but also predicts the actions of the enemy units. Specifically, each time STRATEGIST computes controls, it determines the optimal paths and actions for *all* units (both Blue and Red). Of course, while the controls for Blue UVs can actually be issued to the Blue UVs, the controls computed for the Red units are merely predictions of their actions—which the Red units may or may not follow. Nevertheless, the Blue controls are computed *while giving close consideration to the Red assets' possible actions.*

The potential advantage of STRATEGIST is that it could eliminate the difficulties of human, manual control of UVs by providing, in part, real-time controls produced automatically within a team of UVs and in compliance with high-level guidance and objectives issued by remote human commanders.

STATES AND PAYOFFS

STRATEGIST employs *differential game theory* to arrive at a solution for two opposing forces [2]. Two key concepts are involved: states and payoff function. First, units in two opposing forces have descriptors that describe their *states*. These states change with time. For example, the spatial coordinates of a unit, number of platforms within the unit, amount of weapons and fuel carried by the unit, and so on, all may define the state of the unit. In STRATEGIST, equations that describe the changes of states over time are nonlinear, differential equations. There are some parameters that determine the shape and form of the change that are referred to as coefficients of the differential equations. Of great importance, the change of states is continuous, as opposed to discrete. For example, unit location can be considered with any degree of desired accuracy, to allow a fine-grained tracking of units.

Second, the controller works with a *payoff function* that quantitatively specifies the main mission objectives. Specifically, in its current formulation STRATEGIST assumes that while one of the forces tries to maximize (or minimize) the value of the payoff function, the other tries to minimize (or maximize) the same. This is called a *zero-sum game*, a particular class of game problem where the property of zero-sum invariably holds during the game. STRATEGIST tries to generate the control actions so that opposing units achieve a so-called Nash equilibrium—which guarantees that the two opposing forces have optimal controls (meaning that any deviation from the equilibrium makes the moves of at least one of the forces suboptimal.) STRATEGIST proposes the best possible actions for friendly forces—while also assuming that the enemy will follow the best possible actions from its perspective. The solution, called a saddle point, appears in Figure 12-1.

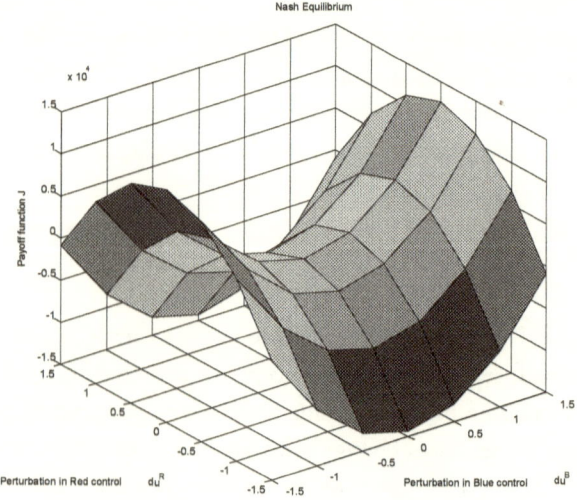

Figure 12-1. The surface depicts change in the value of the payoff function as Blue and Red controls are changed. The saddle point is in the center of the surface where Nash equilibrium is reached.

Suppose the Red commander has only one signal that he can use to control his forces: u^R. Similarly, the Blue commander has control u^B that he can dial up or down. If Red forces are trying to maximize the payoff function, and Blue forces are trying to minimize it, then the saddle point is where *both* Blue and Red commanders are doing their absolute best. Any other point means that either one or both commanders are not entering their controls' best value. Deviations from optimal controls, denoted by du^R and du^B, are plotted in Figure 12-1. Of course, in any real-world situation, a commander has many different ways to influence a battle— that is, he has many knobs and dials, not just one. In that case, the saddle becomes multi-dimensional, impossible to put into a picture but in essence the same.

If a controller were able to compute a Nash equilibrium solution at any given time, the controller would know which controls were optimal for both opposing forces. That's exactly what the controllers in STRATEGIST are programmed to do—to execute an algorithm that finds the Nash solution for a given payoff function.

Even from this brief description, it is clear that this approach uses quite a few simplifications. For example, the fact that the adversary's payoff functions could be asymmetric—in which case the zero-sum approach is an oversimplification, as discussed in Chapter 13. However, the simplifications made in STRATEGIST do have a purpose, allowing fast and computationally feasible generation of control actions. The idea in STRATEGIST is to begin investigation of the game-theoretic mathematical concept as applied to a tactical C^2 environment. In the future, models can be extended to add complexities that parallel the real-world environment.

Since the controller is based on game theory, and separate commands are generated for every Blue unit—and predicted for every Red unit, one can just as easily imagine that there are two separate controllers, one for the Red units and the other for the Blue. Of course, the control actions prescribed for the Red units are strictly predicted optimal control actions—and beyond the Blue forces' control. (The power of having such a separation in offline applications will be discussed further in this chapter.)

Here is a high-level overview of how STRATEGIST works, quickly and accurately:

— Time 0100. STRATEGIST, an automated controller installed on the master UV, receives an electronic situation report, including information about enemy assets and subordinate assets, and the definition of the enemy's mission expressed as the payoff function. (See Figure 12-2.)

— Time 0110. STRATEGIST computes the Nash solution, which includes both trajectories and a plan for weapon engagements (firings) for each subordinate unit for the entire mission, as well as predictions of enemy trajectories and firings.

— Time 0111. Based on the optimal trajectories and firings—not the entire detailed plan, but only for a short time forward—STRATEGIST issues commands to the subordinate units.

— Time 0150. STRATEGIST collects feedback—reports on the actual progress of its subordinate units (probably not exactly what it commanded) as well as the situation and action of the enemy assets (also not exactly what it predicted).

— Time 0202. Using the most recent information, STRATEGIST recomputes the Nash solution.

— Time 0205. STRATEGIST issues a new burst of commands to subordinate units.

BATTLESPACE DYNAMICS

The model formulation in the current prototype implementation of STRATEGIST is very generic. The units participating in combat could be any type—group of UAVs, team of ground vehicles, flotilla of undersea vehicles, and so on—and each may contain any number of entities. For STRATEGIST's planning function, there are no distinctions in the type of units or the ordnance they carry. Thus, while the technology is presented primarily in context of aerial UVs, the concepts apply equally to control of other types of UVs.

A schematic diagram of a model, shown in Figure 12-2, represents a battlespace, where the grid is shown only to specify the coordinates of fixed targets and units. The movement of the other units in the space controlled by STRATEGIST is continuous.

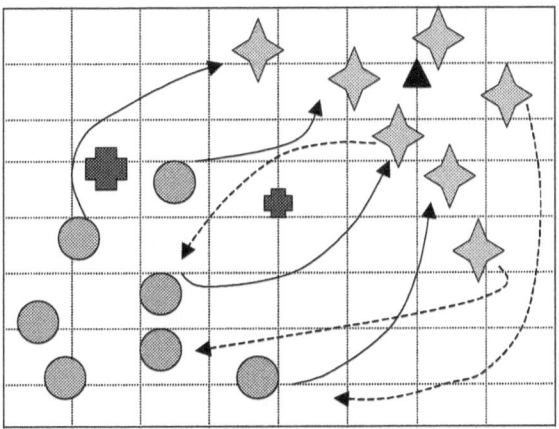

Figure 12-2. The Red and Blue units in a STRATEGIST's scenario.

The figure shows Red units as circles and Blue units as stars; the cross-like figures represent obstacles or targets. Red and Blue units are initially assigned coordinate positions within the grid, and also have assigned strength levels representing the number of platforms and the armament or ordnance they carry.

In the current implementation of STRATEGIST, there are three state descriptors for every unit: location coordinates, number of surviving platforms in the unit, and amount of ammunition within the unit. The list of descriptors can be expanded, of course, to cover various types of UVs, the environment, and the desired degree of modeling fidelity.

The controls exercised over the units include changing coordinates of the unit in the grid, altering the velocity of movement towards or away from an enemy, and changing a unit's firing intensity. (Here, firing intensity is an abstraction used in the current STRATEGIST to describe the extent to which UV's weapons are employed.)

STRATEGIST relies on mathematical models to represent the dynamics of friendly units and their engagement with enemy units. The salient points of these mathematical models are as follows:

— The motion of the units within the region is described by nonlinear differential equations. The equations assume that the units are moving at their rated, nominal velocities.
— Red and Blue units have only one-on-one engagements, and a probabilistic expression determines whether an engagement happens at all.
— While in active engagement the platforms in Red and Blue units undergo attrition. Once again, probabilistic expression determines such losses or attrition dynamics.
— Weapons-expenditure dynamics of the platforms within the units are also expressed in terms of a probabilistic model.

These probabilistic behaviors are used in nonlinear differential equations to describe changes in unit state over time.

GUIDELINES: THE PAYOFF FUNCTION

The choice of the payoff function in differential game theory is critical. As in the original scenario, by sending commands for directional or velocity changes, and for changes in weapons firing intensities, a master UV controls a flock of UVs. Generally, such control decisions result from controller maximizing (or minimizing) the payoff function; in the scenario the payoff function describes the value of destroying various targets and enemy assets, as well as preserving friendly assets. The form of the payoff function would describe, in a mathematical expression, the relative importance of the various units in the environment. The expression may also capture other aspects of the battle; for example, the desired distance between opposing units or desired traversal times. Here, the relative importance of the assets, or other abstractions of combat measures, is governed by the weights or coefficients in the mathematical expression.

For example, in the simplest case, Blue and Red forces each consist of a single unit, with a number of platforms in a unit. The Blue tactical controller needs a weight for the number of platforms in the Blue unit to reflect how important the preservation of assets is to the Blue commander. Similarly, the Blue controller needs a weight for the number of platforms in the Red unit to reflect the importance of how the Red commander perceives the preservation of his assets. In addition, the Red tactical controller needs the corresponding two weights, for the importance attached to Blue attrition.

The current prototype of STRATEGIST uses a quadratic payoff function. It captures the distance between Blue units and their targets, the number of platforms in the Blue units, the distance between Blue units and obstacles in the environment and the weapons spent by the Blue units. Adding all values of these individual terms provides a value for the payoff function as a whole. As an example, a payoff function might read: minimize the distance between the Blue unit 7 and Red target 5 (so that there is a greater chance of engagement); at the same time, maximize the distance between Red target 5 and Red unit 4 (so that the Red unit is unable to protect the Red target); also minimize the attrition of Blue unit 6 while maximizing the attrition of Red unit 3. In STRATEGIST, the goals of Red and Blue forces are assumed to be diametrically opposed—a simplification that is not always accurate—and are reflected in the statement of the payoff function.

The weights associated with Blue forces (representing the importance of different terms in the payoff function) are given to STRATEGIST by the Blue commander, while the weights associated with Red units are based on available intelligence data.

What about providing high-level guidance to the master UV, which uses a technology like STRATEGIST to guide its flock? A human commander in the TCE unit could be the source that provides the payoff function weights to the master UV. Over an interval of time during which such tactical combat lasts, it is expected that such weights will not change a great deal. Thus, the unreliability of the link between TCE and the UVs will not be an overriding concern. If weights were indeed time varying, a human commander at TCE need only issue a short data burst to the master UV with a new set of weights. The master UV would then recompute the new trajectories for the other UVs in the flock—so that the values correspond to a Nash solution. Finally, using data links connecting the master with the flock, new optimal paths and firing intensities would be conveyed to the UVs.

DEALING WITH THE UNEXPECTED

Using the original scenario, STRATEGIST is now onboard the master UV.

The lead UV pair orbits offshore at 5,000 feet, waiting for the F-15E and F-15C formations to join the unmanned armada as it presses northeastward toward the target. Suddenly, the pair receives a burst transmission, establishing contact with the just-airborne TCE. All is going as planned. These formations of UVs convince the Red forces that their threat is only from the incoming Blue UVs and that the Blue forces will employ only them for offense.

This piece of information is also transmitted to the master UV from the TCE.

The lead UV pair computes that they will engage the first of the two enemy radar installations, which are capable of lighting up the incoming fighters, then position themselves to take down any radar that might engage the attacking strikers, fighters, or other UVs. The second UV pair will press into the entry corridor, using all their available weapons to shut down active enemy sites. The third pair will utilize their missiles selectively, only if enemy sites that were meant for destruction were *not* destroyed. The F-15Es and F-15Cs, having already approached the region, are ready to commence bombing the enemy target.

Based on its computations, the master UV also knows the optimal offensive and defensive postures that the Red units could take: the targets are heavily protected by surface-to-air missiles, and fighters will be flying to track down incoming Blue UVs and engage them. Indeed, the entire combat seems almost preprogrammed.

However, realistic combat rarely proceeds as preprogrammed. Enemy actions and other circumstances present unexpected events. How does STRATEGIST deal with unexpected?

Suddenly, Red fighters change their tactics. No longer detecting and engaging the UVs, the Red fighters are now concentrating on incoming Blue manned aircraft. Because the first and third UV pairs are stealthy, they are able to fly through the enemy's radar coverage and escape but the second pair must destroy the Red surface-to-air missiles that could track the Blue manned fighter aircraft.

The situation clearly calls for an urgent Blue reaction. For STRATEGIST, the situation means that the enemy is no longer on the solution path. Such an event is unexpected, but

STRATEGIST has a very simple mechanism to cope with it: the algorithm simply recomputes a new Nash solution. Previously, before the master UV assumed control, it was provided a *threshold value* of acceptable deviation from the optimal solution—for both Red and Blue forces. Therefore, when that threshold was exceeded, the game-theoretic algorithm was triggered, recomputing the Nash solution.

If that's the case, why not preset the deviation threshold at zero, or some other very low value? Would that not minimize the deviation and lead to the best possible results? No, for while a very low value of threshold specification could lead to frequent recomputations—that will indeed enhance the responsiveness of reactions to enemy actions—such recomputations will occur at the expense of increased demand on computational resources.

It is important to note that STRATEGIST deals with unexpected events in exactly the same way it controls an operation that evolves as planned. No special mode of operation is needed. Whether a minor deviation from the course or a major surprise, STRATEGIST is inherently designed for agility in dynamically evolving situations. It is able to handle both situations using exactly the same algorithm and processes.

FILTERING OUT THE ERRORS

An important question can be raised about both inevitable observation errors and deception. To this point, the situation has been described as if all information picked up by sensors, and all intelligence information available to the UVs, or the TCE, was error-free. Such a state is obviously not the case in real-world combat. In particular, the locations and strengths of enemy units are only available as incomplete, missing, or corrupted signals. Therefore, STRATEGIST must deal with observation errors (or, measurement errors in control theory) as well as misleading enemy signals.

STRATEGIST uses *Kalman filtering* to perform *state estimations* for both Blue and Red units [3]. Here, STRATEGIST knows that the state values presented to controllers are not true; instead, they consist of observations corrupted by a variety of factors. Therefore, STRATEGIST must estimate the true state.

A Kalman filter is an algorithm that converts incomplete and corrupted observations—here, of military intelligence reports about enemy units, and of reports from friendly units—into a more accurate estimate of the actual state of the plant. To perform such filtering of corrupted data, a Kalman filter uses both recent prior observations and its own internal model of the plant. For example, a Kalman filter compares two consecutive observations of a Red unit's locations and concludes that the unit seems to be moving much faster than the filter's internal model considers reasonable. The filter then treats the information skeptically.

To filter data, a conventional Kalman filter also uses the knowledge of the control signals that the controller issues to the plant. However, in an adversarial situation a Blue Kalman filter does not usually know the signals that the Red control systems sends to its own units. Thus, the Kalman filter employed in STRATEGIST must operate with only partial inputs: it receives accurate information about Blue control signals, potentially erroneous information about the state of Red units, corrupted information about the locations of Red units, and *no* information about the strength of Red units [4]. The last restriction—the absence of information about the strength of Red units—reflects the fact that, indeed, strength information can be notoriously difficult to obtain, and also gives STRATEGIST a more demanding test.

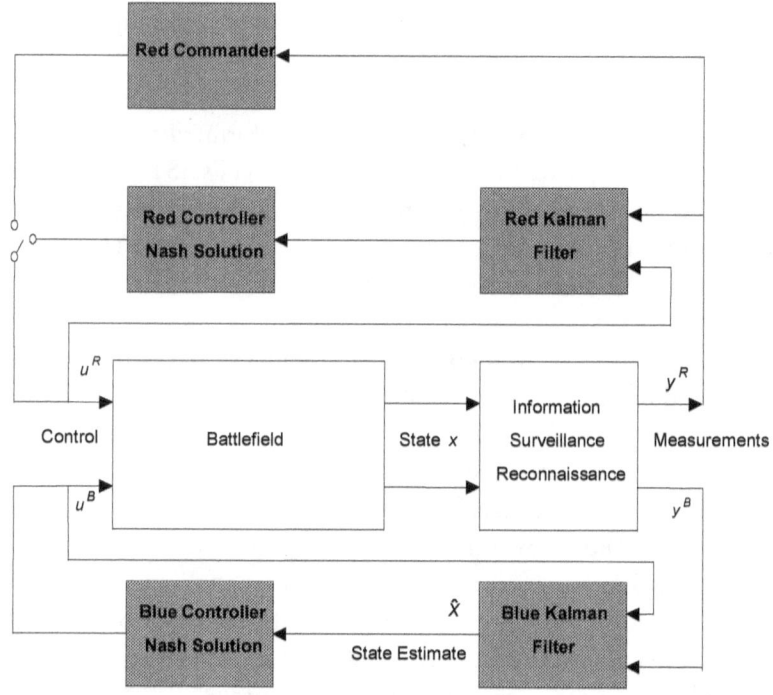

Figure 12-3. STRATEGIST game-theoretic controllers use Kalman filters to convert imperfect observations into more accurate estimates of the true situation.

As an example of a Kalman filter at work, at 1300 hours, the filter using time-update equations estimates a state (the number of platforms in a Red unit) as 10. At 1400 hours, the filter receives feedback from an intelligence report that the actual number of units is four. Naturally, the filter knows that the report could be inaccurate. Using the filter's own internal model of Red and Blue operations and assets, and also measurement-update equations, the filter decides that the most probable number is actually six. Shortly therefore the filter receives another intelligence report, and the process repeats, continuously updating the estimates, gradually getting closer to a true number [5].

A Kalman filter relies on two groups of equations: time-update equations and measurement-update equations. Time-update equations are responsible for projecting forward in time current state and error estimates to obtain a priori estimates for the next step. Measurement-update equations are responsible for feedback; that is, for incorporating a new measurement into a priori estimates to obtain an improved a posteriori estimate. The time-update equations can also be thought of as predictor equations, while the measurement-update equations can be thought of as corrector equations. Here, the time-update projects the current state-estimate ahead in time, while the measurement-update adjusts the projected estimate by an actual measurement at that time.

Moreover, a Kalman filter can deal not only with inaccurate information, but also with missing information as well (within reason, of course). If enough of the rest of the picture is available, a Kalman filter can fill in a missing piece of the puzzle. For example, in the current STRATEGIST prototype, the measurements taken as input by the Blue Kalman filter are those that are relatively easy to obtain: the entire state (locations, strengths, and so on) of the Blue force, location coordinates of Red units, and control signals sent to Blue units. Using these inputs, the Blue Kalman filter continuously adjusts its estimates of Red unit coordinates. However, the filter goes farther, also estimating the missing information that is usually hard to obtain: the strengths (numbers of platforms) of Red units, as well as the control signals sent by Red controller to the Red force.

ZERO-SUM: FINDING A NASH SOLUTION

Computing the Nash solution lies at the heart of the differential, game-theoretic controller. Further, because STRATEGIST needs to produce strategies for real-time, live battle situations, any algorithm that performs this computation must be fast. To this end, STRATEGIST uses a *sequential linear quadratic* (SLQ) algorithm to compute the controls in real time. The method is sequentially iterative, making linear approximations of the nonlinear differential equations governing the motion of the units, and using a quadratic expression for the payoff function. This algorithm, particularly good for the tasks of STRATEGIST, is relatively fast—and therefore suitable for real-time Nash-solution computations by the master UV.

The basic steps in each of the algorithm's iterative cycles to calculate the optimal controls for both Blue and Red are as follows [6]:

1. The master UV STRATEGIST's algorithm starts from both the trial control actions and resulting unit motions—that is, the nominal control and trajectory. Around the nominal control and trajectory the algorithm defines a linearized approximation of the equations representing the unit's motion. The deviations of the unit's motion from the nominal motion are specified by linear differential equations over the entire time interval of combat. For example, if the entire operation is expected to take place over three hours, the algorithm focuses on small changes in the neighborhood of the current nominal trajectory over the entire time interval.

2. The human operator chooses suitable weights for the payoff function required for the tactical situation and provides those weights to the algorithm.

3. Using the payoff function and trial control actions, the algorithm specifies an affine quadratic game—a special type of expression that formally defines the local game to be solved.

4. The algorithm then solves the affine games by integrating *Riccati* differential equations with well-defined terminal conditions. The calculations produce a set of revised control actions for both Red and Blue units. However, these are not yet the optimal actions, but rather control actions significantly better than the original trial control actions.

5. The algorithm then performs an optimality test on the new revised control actions and trajectory. STRATEGIST applies the revised controls to the payoff function, then checks to see if the payoff is still increasing—meaning that it is not yet optimal. This test involves the observed (or expected) state of both Red and Blue units. If the algorithm passes the test, it stops, since it has found the optimal control actions. However, if it fails the test, the algorithm treats the revised control actions and trajectory as new nominal control actions and trajectory. The algorithm then loops back to step 3, repeating the entire process. After several iterations, the algorithm eventually finds control actions that pass the test—that is, they are close enough to optimal control actions.

6. The algorithm has then found the optimal control signals (velocity changes and firing intensities) for all Blue units, as well as predictions for Red units, for the entire time horizon.

BREAK POINTS AND WHAT-IF SCENARIOS

There is also an alternative way of using STRATEGIST, as an offline simulation tool in a decision-support role. An analyst on a commander's staff can simulate combat and find what the end results might be under different conditions, assumptions, and strategies. What-if scenarios may be played out and decisions made regarding initial locations of units, number of platforms, and recommended armaments. As with the online use of STRATEGIST, simulation of what-if scenarios also entails specifying a payoff function that has weights attaching importance to the payoff function terms. A typical scenario might be as follows:

— An analyst inputs the initial locations of the various Red and Blue units, including the number of platforms, weapons, and armament loads. The input also specifies the size of the battlefield grid and the location of obstacles.

— The analyst creates the payoff function with weights (or coefficients) reflecting the importance of the various concerns.

— The analyst invokes STRATEGIST, which simultaneously runs the two tactical controllers—Red and Blue. Invoking STRATEGIST provides Nash solution controls for all units in the simulated battlespace—controls that specify how each unit should be moving in the grid, and the intensity with which it would engage and fire at enemy units.

— STRATEGIST runs the simulation component to play out the effect of the Nash controls. Since the simulation is inherently stochastic (approximating the inherent uncertainty of real combat), the units may not actually follow the paths and actions prescribed by the Nash solution. If so, there will be an automatic recomputation of the Nash solution for both sides.

— To try out different arrangements of friendly and enemy units, the analyst changes the input to reflect such rearrangements, then repeats the same steps to play out the effects. Based on the outcomes of the what-if games, the analyst recommends to the commander suitable courses of action and allocations of assets.

THE QUESTION OF SOURCES

At the tactical level the controller requires the input of very specific, processed sensor information concerning the immediate environment around those platforms engaged in mission execution. The controller also relies on frequent status information from each platform in the mission package. A third key type of information required is fused Intelligence, Surveillance and Reconnaissance (ISR) information concerning the status of opposing forces and targets. This ISR information will be used by TCE or central command to redefine the weights of the payoff function (or, alternatively, qualitative guidance corresponding to such weights).

The weights for the payoff function are an important input for STRATEGIST, and they may be derived by human analysis of information in the ISR database. The ISR database reflects the combined input from the operations and intelligence communities concerning force composition and weapon system characteristics of both potential adversaries and friendly forces. The different categories of weights include those reflecting the win conditions, constraints, and target values. Such information is not normally specified in a format that facilitates use in a tool like STRATEGIST, and therefore would require appropriate means for interpretation and translation, including human-computer interfaces.

Intelligence information concerning the probable courses of action that the opposing forces may adopt is currently available (with varying degrees of uncertainty) in various sources. However, such probable courses of action and enemy objectives are not usually provided in forms that lend themselves to the creation of a clear payoff function. Similarly, friendly objectives for a region or strategic country are often spelled out, but not in a format that lends itself to easy translation into a payoff function for a Nash solution.

At the present time, there is no provision in the controller to handle conflicting reports, although Kalman filters do provide a means of dealing with certain types of inaccuracies. Preprocessing the information flow into the controller may provide a solution, but this process will also limit the usefulness of the controller for real-time operations.

THE BURDEN OF WEIGHTS

Balancing a set of weight coefficients does not seem equal to being in command. Setting weights on the payoff function, and setting thresholds for deviation from Nash solutions, will be a challenging problem for human operators. Since the reliability of the results obtained from the controllers is strongly dependent on the choice of these values, the role of human operators for such situations is critical. Translating objectives to quantified payoff functions is a difficult task. In some cases, several low-level objectives would be combined, and then assessed in terms of a single higher-level objective. Such an audit trail and traceability from higher-level objectives to tactical-level objectives is critical for acceptance of control technology in a distributed environment where several controllers are operating in parallel.

The user interface must allow the user to input and modify the overall win conditions for STRATEGIST, and the format of these conditions must ensure that these critical parameters are understandable by both machine and man. Unfortunately, win conditions are often poorly stated, and decomposing a broad statement into specific parameters to be used by this system may be difficult. For example, an objective of achieving air superiority may be associated with a win condition that specifies enemy aircraft will not fly missions in the defined area. How would that win condition be translated into specific tasks? Do we strike all airfields capable of launching missions? Do we shoot the enemy aircraft down once they are airborne? A human must resolve such key issues before guidance is passed on to the controller.

Apart from the issues relating to weightings in the payoff function, there is another important interface issue that needs to be solved. This has to do with target valuation. The controller operates at a sufficiently low level in the overall command hierarchy that it only anticipates receiving a target list with a consistent set of values. This valued list supports the cost calculations involved in the internal controller model. The process and means used to calculate the target value are a significant issue for resolution.

SUMMARY

With STRATEGIST, a leader in a team of UVs would continually be able to compute optimal controls, issuing them to subordinate team members. Meanwhile, a remote human operator would need only to issue high-level guidance to the leading UV, thereby eliminating both the current difficulties of manual control of UVs, as well as the excessive reliance on long-distance communication links. Alternatively, STRATEGIST could be used to provide intelligent controls to simulated platforms or units for what-if, decision-support wargaming.

STRATEGIST approximates combat of multiple units or platforms as a zero-sum game problem—that is, a game in which opponents have exactly diametrically opposing objectives. While the zero-sum simplification may not be entirely realistic, it does allow the use of very efficient, real-time computations. At every moment, the fast SLQ algorithm computes the best possible sequence of actions, from the beginning until the end of the game, for both opponents (assuming that each one does his best). This computation is the so-called Nash solution.

The model is based on nonlinear differential equations, and the objective (or payoff) function is a combination of assets and target values. In spite of many simplifications used in the current prototype—for example, constant velocities, one-on-one engagements, and intensity of firing—the approach does not preclude extensions to higher-fidelity models.

The Nash solution is recomputed in real time, and whenever the situation deviates from recent expectations. Hence the approach is well suited to handling unexpected events with great agility.

Recognizing the inevitably erroneous and incomplete nature of available observations and intelligence, Kalman filters perform state estimation—that is, the conversion of imperfect and possibly deceptive observations into a more likely true situation. Within reason, STRATEGIST can correct erroneous Information, and even fill in missing pieces—for example, information about enemy strength.

CHAPTER 13

WHEN OBJECTIVES DIFFER

In most applications of game theoretic methods, technologists work with the assumption that both friendly and enemy forces are concentrating on the *same* set of objectives. In many cases, however, there is a multiplicity of objectives.

In the previous chapter, which discussed game theory in military applications, combat scenarios had multiple objectives that were combined into a *single objective function* over which a zero-sum game was played. In this chapter, situations are presented in which game theory is applied in a battle with multiple objectives—but where these objectives need not, and cannot, be coalesced into a single function. Here, the technology handles these objectives independently.

One of the most significant problems in adversarial situations occurs when opponents' objectives are significantly different. Imagine a battle in which Blue forces are trying to achieve a certain goal—securing an area within Red territory. Red forces, however, are concentrating on a different goal: not concerned about holding territory, they are instead targeting a power station within Blue's borders. Technically speaking, the two objectives are orthogonal: while certainly, Red and Blue forces are working against one another, their immediate objectives are quite different, and there is no direct correlation between the two. How does a Blue commander develop the best strategy in such a situation?

A hypothetical Blue order might say that the advance of Red troops must be halted before they cross the border into Blueland. While

Blue forces know that Red will try to advance towards some strategic locations in Blueland, available military intelligence information is unreliable. Therefore, the Blue commander cannot be sure whether incoming Red troops are intending to capture the oil refinery or target the ammunition depot, 40 miles north. The Blueland commander lacks the forces to mount a strong defense in both locations and therefore must prioritize his efforts. One area would become the *main effort* for his defensive operations, while the other would be an *economy of force*, or secondary, effort. In determining which location would receive the strongest defense, the refinery is extremely important to Blueland; it is their main source of economic strength, they want to protect it with whatever means they have. Clearly, there would be a mismatch between the objectives of Blue and Red, if the latter were to decide to attack the ammunition depot. Does a commander have the tools and technologies that can let him analyze such situations?

Furthermore, positing a situation fraught with even more complications, there seems to be a divergence of objectives within Blueland, because while the Blue troops are interested in protecting the ammunition depot, the local civilians want to make sure that the refinery is defended. Their reasons are easy to see. In addition, the Blueland central government has issued a clear directive to the Blue forces that the refinery must be protected. Therefore, a commander might ask, how should one provide guidance to subordinate forces regarding a preferred operational objective— when there is also a multiplicity of objectives within different segments of the same party to a conflict? How would a Blueland commander optimize his decisions in such a situation?

In Chapter 12, the techniques of game theory were applied to military situations in which there were no such mismatches of objectives between the two forces. However, a more realistic situation includes a significant mismatch. In addition to dealing with such a complexity, there is also the question of influencing subordinate and allied units all at the same time.

It is for precisely these complex challenges that this chapter presents new developments by a team of control technologists at Ohio State University, in Columbus, Ohio, with their collaborators at the University of Pittsburgh, in Pittsburgh, Pennsylvania. These researchers have developed game theoretic controllers that are potentially applicable to command and control scenarios with multiple and different objectives [1]. Although the current work is limited to fairly small and simple scenarios, it is possible that ongoing efforts can prove useful in the types of situations described above. Such controllers generate strategies for friendly forces, and predict strategies for adversarial forces, based on *different* objectives specified (or assumed, in case of the enemy) for each side. The controllers also address the notion of *teams* of units, or agents, that may be arranged hierarchically on two opposing sides [2]. Hence, the technology is referred to as *Stable Multi-Agent Strategies in Hostile Environments,* or SMASHER.

SMASHER is a controller in a battle simulation, generating and issuing control signals to *simulated* Red and Blue units. A tantalizing possibility exists that a technology, similar in approach to SMASHER, might also control actual combatants. Indeed, such a controller might be able to provide control signals to Blue UVs (or suggestions to manned units and platforms) while simultaneously predicting the most likely Red moves. Because of the extensive computation requirements, this is a daring concept for the future. In this chapter, SMASHER uses a simulation tool to help perform its role as a decision-aid.

THE CASE OF MISMATCHED MULTIPLE OBJECTIVES

Recapitulating elements of game theory, as presented in Chapter 12, two opposing forces have objective functions that they try to maximize and minimize [3]. The functions may be specified as complex mathematical expressions describing such considerations as capture of targets, destruction of enemy assets, preservation of friendly forces, psychological and economical effects, and so on. In many tactical situations, they could be modeled, as a first approximation, as a combination of terms depicting the importance of the assets employed in combat. This situation, then, implies that the multiple objectives are captured in a single objective *function,* and that both sides are interested in the same set of objectives.

Further, usually a game theoretic formulation is given as a zero-sum game, in which the *same* objective function is maximized by one side and minimized by the other. The solution to such a zero-sum game provides *optimal* solutions—solutions that guarantee the best possible actions for the two opposing forces. Therefore, if one side chooses to perform any other action (not provided by the solution), the value of the objective function, from the perspective of that side, is less than optimal. This is called a Nash solution [4].

The problem with such a zero-sum approach, however, is that it tends to be unrealistic in situations where the objectives of the opposing sides are mismatched. In a situation of mismatched objectives, an analytical tool must model the multiple objectives of both sides not as one objective function but instead as two (or more) separate and different functions. Because SMASHER allows Nash solutions to be obtained over an arbitrary number of independent objective functions, it can be said to provide non-zero-sum Nash solutions.

The power of such a mechanism is that it allows handling multiple and potentially orthogonal objectives, as in the Blueland scenario. SMASHER is therefore different from most game theoretic techniques in that it allows specification of multiple, potentially mismatching objectives, as opposed to a single objective in which a zero-sum game is played.

THE CASE OF LEADERS AND FOLLOWERS

Another aspect of SMASHER is the *leader-follower* game theoretic strategy [5]. In this control method, it is possible to designate a leader among several team players who steers the outcome to a desired point. Here, the leader acts by stating terms at the game's onset: if the followers choose certain actions, as opposed to other actions, then the leader in turn will choose actions that will benefit the followers. Thus, the followers follow the leader's objectives by choosing those alternatives that would maximize (or minimize) their own objectives. The solution obtained in such a case is called the Stackelberg solution, and it represents the best actions for the leader and the followers [6].

Since the leader and followers are on the same side of the conflict—that is, their overall goals are the same—this is called a leader-follower game in a cooperative situation [7]. Thus, SMASHER models two leader-follower games: one Blue, one Red [8].

Contrasted with these cooperative games are non-cooperative games—or those between adversarial units whose respective objectives are opposed to each other. In non-cooperative situations, SMASHER employs non-zero-sum Nash games, as discussed above.

The inspiration for both game theoretic techniques—Nash and Stackelberg—is derived in part from economic models, finding the best strategies in competitive environments. When there is no cooperation between two or more decision-makers, and there are no means to enforce agreements, a Nash solution is well warranted—since both opposing forces achieve the best outcome in the given circumstances. The leader-follower approach warrants the Stackelberg solution, and is therefore suitable for developing strategies that influence and guide the subordinates.

CONTROLLER OF SIMULATED BATTLES

As the Blueland scenario suggests, SMASHER's main function is its use as a *decision aid* for developing strategies and analyzing courses of action in a battle with multiple and mismatching objectives.

SMASHER generates optimal controls, which manifest themselves as signals to respective simulated units, specifying when and where to move, how much ammunition to use against which adversarial unit, and so on. At any given point in time, control actions generated by SMASHER depend on the value of both the different objective functions and the current *state*—that is, the situation of the units.

Apart from the controllers that actually calculate the best controls to exert, SMASHER has a *simulation* module that can play out the effects of generated controls. Figure 13-1 gives a sketch of how a military enterprise for C² is modeled in SMASHER.

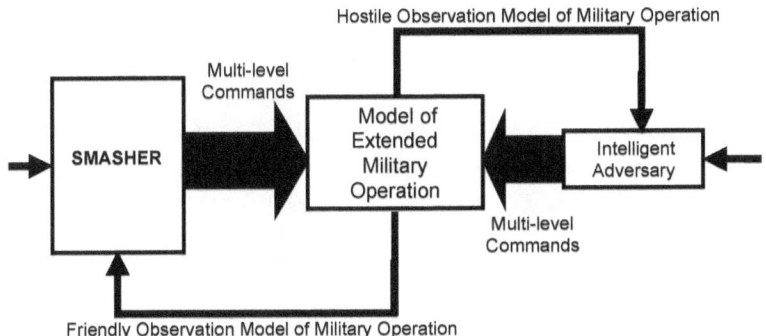

Figure 13-1. This model of a military enterprise includes SMASHER and a model of an intelligent adversary.

As in other game theoretic techniques, SMASHER determines the best controls to exert for both Red and Blue units: Red and Blue controllers within SMASHER independently arrive at strategies. Based on the objective functions they have to maximize, these two independent controllers direct the units' movements on the grid, decide which opposing unit to engage, and choose when and how much ammunition to use.

A significant point about the SMASHER algorithm is that the controller has to have knowledge of all the participating units— which is possible in a simulated situation. However, if SMASHER were extended to control real combat entities, such as UVs, complete knowledge of all participating units, especially Red units, would be clearly impossible. In that case, SMASHER would rely on the best available military intelligence.

THE DYNAMICS OF RED AND BLUE

In effect, SMASHER models both Red and Blue high-level commanders. Following a Stackelberg strategy, a high-level commander on either side influences the moves and progression of the units that he commands [9]. By using Nash strategies, units at peer levels achieve new states by accepting optimal commands from their respective controllers.

The environment set up in the current prototype model is a rectangular geographic area arranged as a grid of finite size, with grid locations described by x and y coordinates. A schematic diagram, including lower-level participants, is shown in Figure 13-2.

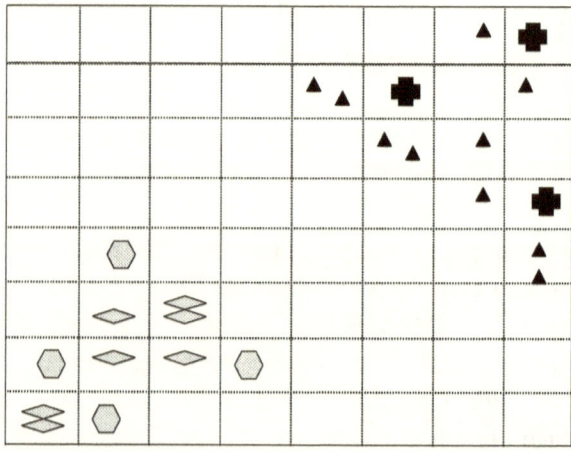

Figure 13-2. The rectangular grid represents the environment in which SMASHER controllers execute controls over their respective units.

The model formulation in SMASHER can be generic, in the sense that the units participating in the combat can be of any type, and the number of platforms (actual fighting entities) that the units contain can be varied. For example, comparing the Blueland battle with Figure 13-2, it could be that the Red forces are triangles, moving aggressively, perhaps as armored formations. The hexagons could then represent the Blue oil refinery and ammunition depot, with the diamonds near the Blue installations those Blue units that would try to thwart the Red advance.

Here, four discrete descriptors define the state of each unit: location (x coordinate), location (y coordinate), number of platforms in the unit, and number of weapons per platform. (Of course, some descriptors may not apply to all units; for example, number of weapons will not apply to the refinery.)

A key issue is how to model the *plant* dynamics; that is, movements and other changes in the states of Red and Blue units. As the units in the environment move, experience attrition, and expend their ammunition, SMASHER must capture such evolution. In other words, mathematical expressions are needed to model unit dynamics; that is, at a given moment how the state of a unit depends on the state of that (and other) units at preceding times. In the current implementation, SMASHER uses fairly simple, approximate expressions for describing plant dynamics—which may suffice under many conditions, although a more extensive, higher-fidelity model would be required for realistic applications.

SMASHER models dynamics of Red and Blue units with discrete-time, nonlinear equations. Here, discrete time means that changes in unit states only happen at distinct instants. Nonlinear means that the rate of change of any state variable (for example, ammunition used) is not linear or proportional to time spent during battle. Therefore, depending on circumstances, the rate could be small at one instant and large at the next. Further, to reflect the non-deterministic nature of combat, there are also some random elements in the model:

— In the course of the evolution (that is, movements and

engagements) of the battle, a probabilistic expression is used to determine whether an engagement between two opposing units will happen or not. In a sense, at each time instance the model rolls dice to assign engagements of Blue and Red units—although the dice are carefully biased to reflect the realistic likelihood of an engagement actually occurring.

— Further, as the units' states evolve in time, their strength will also change, represented by the changes in the number of platforms in all Red and Blue units. Changes in strengths (due largely to attrition) are also modeled probabilistically— where a probabilistic expression determines how much attrition happens to each unit at each time instance. Kill probabilities are used as parameters in these expressions.

The nature of these probabilistic expressions could also capture the mismatch of battle objectives—for example, those expressions governing engagements could be biased to capture shifting goals.

The benefit of modeling the dynamics in such a manner is its relative simplicity: what matters in the evolution of dynamics is the current situation of the units, and the *most recent controls that have been exercised on them*. However, the fidelity of the model can be increased in the future as needed, without invalidating the basic ideas of SMASHER's approach.

SMASHER controllers generate signals executed in the simulation for every state of Blue and Red units. For example, those controls sent for a unit are: relocate to new coordinates (that is, specifying the x and y locations on the grid), the quantity of weapons to use in firing (for example, salvo size), and the targets to engage.

MULTIPLE OBJECTIVE FUNCTIONS

How could SMASHER's models specify that Red forces are to take an aggressive stance or that Blue should thwart the enemy advance by blocking off the routes to the oil refinery or the ammunition depot? The answer lies in the way that objective functions—the mathematical representation of the importance a commander attaches to various assets, targets, and so on—are specified. For example, commanders on both sides are interested in minimizing attrition of their assets and maximizing those of the adversary. In the leader-follower game, SMASHER's model posits a high-level commander controlling multiple subordinate Blue or Red units—and specifies an objective function over which a Stackelberg solution can be obtained. Such a solution directs the subordinate units to follow the commander and maximize preservation of their own assets. The objective function thus specified for this situation mathematically expresses the commander's preferences. Quite often, then, the function is simply a sum of friendly assets' strengths minus the adversarial assets' strengths. To express the importance of the unit or asset, weights or coefficients are applied to the number of platforms and weapons. Thus, the choices of these weights in the objective function reflect the commander's preferences and priorities.

In a function that replicates bottom-up, as opposed to top-down objective function planning, low-level units have their own objective functions specified in a similar manner, representing the conflict with their peers on the opposing side. This is clearly a non-cooperative situation, and, hence, in SMASHER, a Nash game is specified among opposing low-level units. Once again, weights or coefficients attached to each term in the objective function reflect a term's relative importance.

To use SMASHER, a staff analyst creates the models, specifying the number of platforms in each unit and the amount of ammunition each carries. To specify an aggressive stance of Red forces, for example, the analyst would choose large negative weights for the important targeted Blue assets. In an effort to maximize the objective function, the controller for the Red forces will give preference to those actions that go against Blue's objectives. By choosing appropriate weights for different terms in the objective functions, then, SMASHER can be used to program different intentions of either enemy or friendly forces.

For example, to model the Blueland situation the weights in Blue's objective function have large negative values for the oil refinery and ammunition depot. In using SMASHER as a decision aid, however, an analyst can alter the scenario by introducing different types of objective functions, then determining outcomes in each case. If it is indeed decided to model Red intent to capture the refinery more than destroying the ammunition depot, then the weights in the objective functions of the Red low-level units would be so programmed. Similarly, for Blue's defensive efforts against Red, if protecting the oil refinery is more important than saving the ammunition depot, then the weights attached to each asset are set to reflect that preference, and the Blue units would follow the desired objective.

BOTH NASH AND STACKELBERG

In a cooperative situation, SMASHER uses the Stackelberg theory of games, in which there is one leader and one or more followers. By choosing a suitable objective function, the followers are induced to perform the desired actions of the leader. In a non-cooperative situation, SMASHER relies on the Nash theory, which determines optimal strategies for two opponents. Now, how exactly is the game played? The key features of the algorithm employed to determine optimal control actions for each unit in the battle include:

— Objective functions are specified by the analyst for high-level and low-level 'units. These objective functions represent the importance, or cost, of targets, units, etc.
— Costs are made available by the analyst for every entity in the battle and all possible control actions that each could be asked to perform. These costs are stored in memory, in a

tabular or matrix form, for use by the algorithm. As an illustration, if a Red unit can perform three actions, and a Blue unit can perform three actions, the costs derived for each permutation of actions are recorded as pairs in a 3x3 matrix. Accordingly, the matrix is exhaustively specified for every possible combination of actions.

— The algorithm proceeds in discrete time steps. At every time instance, the algorithm performs two different kinds of searches: for non-cooperative games (those between peer-level opponents) it searches for the minimal (or maximal, based on how the objective function is specified) value for the cost pairs—as in a Nash solution. For cooperative situations, where a high-level unit is the designated leader and low-level units follow it, the algorithm searches for the mutually accepted minimal (or maximal) cost—as in a Stackelberg solution.

The accompanying simulation module in SMASHER is a simple Monte Carlo simulator that helps perform the sensitivity analyses. In the simulation, game evolution proceeds forward in time; as the states of the game change probabilistically, controls—based on Nash and Stackelberg solutionss calculated at a given time instance—are exerted. Multiple runs are performed, and statistics are gathered. Typically, as some parameters (kill probabilities in expressions for attrition, for example, or weights in the expressions for objective functions) are perturbed, outcome variations (average number of remaining Blue units, for example) are recorded to make inferences about sensitivities.

THE ROAD TO BLUELAND

By running the simulation multiple times, the analyst on the commander's staff can perform what-if analyses. For example, he may try out different initial locations of Blue and Red units, and then observe the outcomes in the simulation. The analyst can also choose different sets of objectives to get an idea of possible outcomes.

In the Blueland battle, SMASHER can, for example, determine the battle outcome with two different Red objectives: one in which it would target the oil refinery, and the other targeting the ammunition depot. SMASHER's controllers would execute the battle, giving priority to the appropriate targets in each case. The analyst would then analyze the outcome and make recommendations to the commander.

As previously mentioned in Chapter 11, it's also important to note sensitivity analyses; that is, the what-if studies of how sensitive combat outcomes are to changes in various parameters. Here, the simulation environment provides a means to assess the vulnerabilities of both Red and Blue forces. As a simple example, a set of simulations may examine the remaining number of Blue platforms after a round of combat. If perturbations around a given value of kill probability do not affect the outcome, then Blue forces have some robustness in their strategy.

To illustrate a typical SMASHER usage, here are the steps that an analyst would follow in the Blueland battle (with a high priority placed on protecting the oil refinery and Red troops stopped from advancing very far into Blueland):

1. The analyst defines the high-level objective function for all Blue forces, including a large positive weight on the oil refinery and a smaller weight on the ammunition depot. Since the objective function may capture several other aspects of the battle, a large weight is also assigned to the number of time steps required to annihilate Red forces. In this manner, the requirement to achieve decisive results rapidly is also captured, at least approximately.

2. The analyst defines the high-level objective function for the Red forces, based on the best available military intelligence and assessments of enemy intent (in this respect, see Chapter 4, which describes a technique for determining enemy

intent). Assuming enemy intent is to capture the ammunition depot, a large negative weight placed on the depot by the analyst reflects the importance of this intent. SMASHER's Red controller algorithm will then automatically attempt to maximize the achievement of this objective for the Red forces.

3. The analyst defines low-level objective functions for Red and Blue forces; that is, objective functions for each subordinate unit on both sides. To reflect the differences in intents and objectives of various units, multiple objective functions for units on both sides are specified.

4. As part of the model of plant dynamics contained in the SMASHER database, the probabilities of kill and attrition are already pre-specified. However, the analyst modifies some of these numbers to reflect latest experiences and most recent estimates.

5. The analyst invokes SMASHER, which simulates an entire battle in which both Red and Blue units are driven by signals produced by controllers within SMASHER. Then, using different weights in the objective functions to reflect different possible intents, the analyst performs additional what-if analyses. After each simulation, the analyst observes the outcomes, gathering such statistics as the mean and variance of outcomes (for example, Blue attrition).

6. The analyst also decides to perform sensitivity analyses by varying the kill probabilities or the weights in the objective functions, then entering these parameters and executing multiple simulation runs to harvest sensitivity information.

In this case, the analyst tries out various deployment schemes, entering the different weights on the objective functions for the simulated Blue commander. Based on an assessment of the upcoming battle, the analyst assigns great importance to the Blueland refinery and less to the ammunition depot. Appropriate number and type of platforms, and strength values for Red attacking units and Blue defending units, are also assigned. Weights of the objective functions for the simulated Red commander are also assigned, in such a manner that Red units will fight aggressively for the capture of the oil refinery. As the SMASHER simulation runs with this set up, the analyst may determine, for example, that the Red force threat is not particularly significant.

Here, SMASHER helps arrive at useful conclusions in a systematic and quantitative manner. Of particular importance is that the actions pursued by both Blue and Red units, and their simulated commanders, were rigorously derived from well-defined intents, objectives, and preferences. Unlike a typical simulation or wargaming system, SMASHER's simulation is accomplished without relying on preconceived, hard-coded rules of typical or doctrinal behavior of forces. At least as important is the fact that none of this simulation is based on any assumption about symmetry of opponents' objectives.

Another intriguing possibility would be to use Stackelberg theory to influence the enemy in such a manner as to gain an advantageous position. However, SMASHER's capabilities in such a role have not yet been explored.

FOLLOW THE LEADER

Because the SMASHER technology is capable of dealing with mismatching objectives, doing so with rather succinct models, the technology potentially represents a new technique whereby a commander and his staff could analyze operations that have a multitude of objectives with complex interdependencies. Whereas most game theoretic technologies have relied on zero-sum games with a single objective function, SMASHER presents a more realistic departure.

Coupling the Stackelberg leader-follower concept with multiple objectives specifications creates a conceptually easy way to capture the complexities of a battle with mismatched objectives, such as the Blueland example. Here, SMASHER's advantage over alternative approaches is notable when specifying various intents in battle. While formulating mismatching objective functions is not a trivial task for the analyst, at least its conceptualization is relatively simple.

It should also be noted that the SMASHER technology is most suited to scenarios in which the number of model entities or participants is fairly small. If the number of entities is large, the algorithm quickly runs into computational problems. Even for small situations, listing out permutations of all costs in the matrices can becomes an expensive computational task. Thus, the use of SMASHER is probably best suited for higher, less detailed levels of analysis—where the number of states and entities are necessarily aggregated and hence fewer in number.

REAL WORLD, REAL TIME

The underlying model of SMASHER's plant has to be specific for every battle situation. In particular, a number of parameters and data must be input into the tool as part of *initialization*:

— The analyst must enter the details of all units on both Red and Blue sides. For different rounds of analyses, different configurations of Red and Blue units may have to be entered. As a solution, the tool may be interfaced with a database from which the analyst may select this information for SMASHER.

— The analyst must specify initial location coordinates of each unit on the grid, number of platforms each unit has, and weapons and ammunition in each platform.

— The analyst must enter various parameters, such as

weighting parameters (indicating the importance of Red and Blue resources) and kill probabilities.

For larger scenarios, data entry could be a relatively long and cumbersome process that should be supported by adequate computer-user interfaces. It should be hoped, however, that such a set-up is in all likelihood significantly shorter than in the case of alternative simulation technologies, particularly because, unlike conventional simulation technologies, SMASHER does not require modeling rules of unit actions and general behavior.

Apart from initialization data and parameters, which are read in as initial inputs, Red and Blue controllers in the simulation environment must receive other run-time inputs as well. In the simulation, evolving dynamics of Red and Blue units are observed perfectly, implying that (due to stochastic evolution in the simulation) *new state information* is available accurately and without delay. Although such a situation is an unrealistic representation of a real battle, SMASHER's capabilities can also be augmented to incorporate imperfect observations.

If SMASHER is used as a real-time decision-support tool—that is, if simulations are performed to determine adjustments of real-unit tasking as the battle progresses—to reflect any new battle status as discerned from military information, the tool must be interfaced with an online feed that periodically provides new values of the *parameters* (such as weights in objective functions).

As the simulation progresses, an analyst performing what-if analyses may have to see the *control signals* as they are computed at each time instant. Therefore, interfaces to display this information are also required. For small scenarios, such interfaces can be provided with histograms; as has already been implemented in SMASHER. For larger scenarios, as the simulation progresses, there must be a means to provide aggregated information to the analyst.

When multiple runs of the simulation are performed, SMASHER collects *statistical data*. The tool then outputs result data that may, in turn, be fed into packages to render these results graphically. From such graphic data, sensitivity analyses may be performed. In its current implementation, for example, SMASHER provides three-dimensional plots for sensitivity.

MAN AND MACHINE

In performing what-if analyses with SMASHER, the analyst will want to change the different weights in objective functions. With a multiplicity of objective functions, however, a human analyst may find it difficult to determine which weight to change. Further, it is hard for an analyst to perceive the effects of changing weights: having observed an effect, the analyst may not be able to isolate the cause. In general, when a large array of weights and parameters is to be specified, it might be difficult for an analyst to set their quantitative values.

To contend with the difficulty of having to specify numerical weights within objective functions, some other means of specification may be required. In this connection, there are interesting developments in *ordinal game theory*. In this approach, the analysts need only specify the *relative importance* of the terms in the objective function. The cost matrix that is then built includes only ordering information. The algorithm can work from this ordinal information in much the same way that it currently does with cardinal information in SMASHER's cost matrix. Such a method might significantly enhance the usability of SMASHER technology.

SUMMARY

Although focusing on a similar problem as in the previous chapter, this chapter searches for ways to eliminate two major simplifications. First, a solution is proposed that does not assume opponents have exactly diametrically opposed—that is, zero-sum—objectives. Second, the solution recognizes multiple, not-entirely-identical, objectives that exist at different echelons and organizations of any given side. Elimination of these two simplifying assumptions leads to a more complex problem.

To find a Nash solution for this more complicated non-zero-sum game, SMASHER uses a discrete-time, nonlinear model for the dynamics of the units and a grid representation of the battlespace. In current prototype, SMASHER introduces such simplifications as allowing engagements only if opposing units are in the same square of the grid. The objective function, as in previous chapter, consists of values of targets and assets, with weights, but with potentially different weights for each opponent. SMASHER's approach could lead to computational explosion for a scenario with large number of units.

SMASHER also models the relations between superior and subordinate echelons as a leader-follower game theoretic problem—a kind of cooperative game where leader and followers strive to maximize their respective (but not identical) objectives by doing the right things for each other. The so-called Stackelberg solution for such a problem represents the optimal actions for both leader and followers. Although not entirely representative of real humans' or organizations' behavior, this solution is a useful approximation, an aid to deciding how to influence other friendly entities.

The near-term use of SMASHER would likely be for computing control signals to simulated platforms or units, and for what-if, Monte Carlo wargaming. However, it is not impossible that with further advances SMASHER may apply to real-time control of UVs.

CHAPTER 14

BATTLE AS A BOARD GAME

Warfare and strategy board games have been popular for centuries, offering not only recreation, but also the ability to replay historic events, teach logic and strategy, and inspire intellectual competition. Historically, boards have been made of stone, wood, and cardboard; now, there are virtual representations in cyberspace. In addition, board games have evolved from simple two-player games, like checkers, to complex multi-player games, like those involving hundreds of Internet players. With such a rich history, and diversity, the question arises: could war board games be used as a basis for C^2 systems?

In all board games, the board represents the physical area of confrontation. Some boards may have geometric patterns, such as square grids (chess) or hexagons (Chinese checkers) while other boards are map-based (Risk™). Indeed, the board's attributes help define and bound the actions that can occur in the game, including constrained paths, obstacles, and other limits. Similarly, C^2 systems use maps with rectangular grids that can also include terrain features, political boundaries, and other attributes that affect movements. In this way, the C^2 playing surface has attributes similar to those in board games.

Board games also have pieces that represent the actors. Chess, for example, has six types of pieces, each with a clearly defined role and rule of movement. Modern war games have pieces, too, that can represent a variety of military entities, from an individual soldier to a full corps. Similarly, in a C^2 system, forces at any and all levels of the hierarchy can be shown symbolically on maps; this way, playing pieces can represent forces as they would in a board game.

Taking the game a step further, moves then represent critical events. While in traditional board games moves are taken in-turn, now, in many Internet-based games, multiple pieces and multiple players can make concurrent moves. Similarly, in a C^2 system, moves can also be performed concurrently—effectively putting C^2 on the cutting edge of contemporary game practice.

For all games, engagements represent the outcome of moves; in board games, some engagements are deterministic. For example, in chess the outcome is clear when the black bishop moves into the space occupied by the white rook. In other board games, however, engagements are random (stochastic). For example, in Risk dice rolls determine the moves into an opponent's space. Similarly,   because   in C^2 systems engagements can be modeled deterministically (with probability tables) or stochastically (randomly), play actions can be modeled in the same way that they occur in a board game.

Therefore, it would appear that war board games have the necessary fundamentals to support a C^2 system. At this point, it might seem that to go to the next step, to create a decision-support tool, all one would need to do would be to add some computational power to automate the game-playing process.

However, it's not that easy. If a computer is to coordinate multiple pieces in a real-time game board representation of an engagement, there are at least two requirements. First, the computer must have a computationally feasible algorithm to generate good strategies. Second, the computer must also have the ability to react quickly to adversarial moves so that it can enact the strategy. To satisfy these two rigorous requirements, any technology must provide a commander with two vital assurances: that it possesses an effective method (based on sound expert strategy) for planning the movement of units or platforms, and that it contains some lower-level automatic feedback routines for the units or platforms to use to as they engage the enemy.

A MOVER AND A SHAKER

Researchers at the Rockwell Science Center, Thousand Oaks, California; Stilman Advanced Strategies, Denver, Colorado; and Wayne State University, Detroit, Michigan, are developing Mover and Shaker, two technologies to combine intelligent reasoning with fast feedback control. Linguistic Geometry (LG), the basis for Mover, dynamically generates mission strategy [1]. Finite State Machines with Parameters (FSMwP), the heart of Shaker, uses DES-based hierarchical modeling and control framework to provide fast, low-level, event-driven control [2].

Apropos of using LG for creating mission strategy, Professor Boris Stilman, the originator of LG, states, "There are many problems where human expert skills in reasoning about complex multi-agent systems are incomparably higher than the level of modern computing systems with respect to complexity reduction. Though there is no grandmaster in combat simulation or robot control, in the game of chess the human grandmasters have achieved amazing results in search reduction. One of the unique features of the LG approach is the formalization and utilization of search heuristics developed by highly-skilled human experts (such as chess grandmasters) [3]."

Briefly, Mover builds strategies that specify how to move pieces in response to adversarial moves on the board. Constructing local goals for every piece at each time step, Mover also generates sequences of moves that allow each piece to achieve its next local goal. Here, local goals are usually equated with desirable events—reaching a certain location, for example, or destroying a certain enemy piece.

Shaker, to the contrary, produces lower-level, real-time control signals for the real-world units or group of real-world units represented by each game piece. Responding to specific events in the battlespace—units being attacked or destroyed, for example, or undergoing mode-changes—and based on the observation of specific events, Shaker provides control actions, such as evade, attack, and so on, to each member of the group. Importantly, Shaker also coordinates the units or platforms in the group. For example, when one unit of the group observes an event, Shaker provides control actions to all members of the group.

Mover and Shaker combine in a hierarchical fashion, with Mover residing at the higher level and performing long-term strategy generation, while Shaker resides at a lower level, performing fast local feedback control. This way, Mover's strategic game board moves are translated to specifications for Shaker, which in turn synthesizes feedback control. Such a two-tiered approach partially alleviates the combinatorial explosion associated with Mover's strategy generation. First, because Mover only deals with pieces that might represent groups of units, it reduces its total number of pieces, and also the number of choices for each move. Second, because Mover does not explicitly deal with a unit's mode (cruising, evading, and so on) and is only concerned with the configuration of the game board—each piece's actual location and orientation—it therefore does not have to carry excess computational baggage.

Nevertheless, hostile counteraction occurs at both levels of this two-tier approach. During dynamic strategy generation, Mover takes into account potential enemy moves and strategies. At the same time, using policies that ensure safety, or avoidance of pre-defined, undesirable states, Shaker counters adversarial events within the discrete model for each unit—getting out of cruise mode, for example, when being tracked by enemy radar.

THE BEST OF BOTH WORLDS

The primary advantage of such a two-tiered approach is that it allows higher-level intelligent control and coordination of a complex, multi-agent battlespace while still allowing real-time reaction to lower-level enemy action. Overall, the advantage of LG technology over other strategy-generation methods, such as brute-force search with minimal pruning, or dynamic programming (see Chapter 2), is that LG remains tractable—that is, computationally feasible—as the game board grows in size. Here, a standard brute-force search is replaced by a construction process that, based on enemy moves, generates sets of future move sequences. Moreover, LG technology also has an advantage over search-based algorithms because, by illustrating current proposed movements as responses to possible enemy movements, it supplies relatively intuitive explanations for suggested future movements.

The primary benefit of FSMwP is that it allows fast feedback control and coordination during implementation of LG moves. Along the same lines, the advantage over other finite state machine models is that, by utilizing continuous/discrete parameters to represent such continuous/discrete resources as fuel levels and ammunition counts, FSMwP avoids state-space explosion problems. Such discrete parameters thereby separate the representation of resources from that of behavior, which appears in compact form in the graphical representation of the FSM. In addition, the FSMwP also permits multi-level hierarchical construction, where a single model represents groups of units or platforms.

A secondary advantage of the FSMwP approach is that a mission's formalized and low-cost data record can be saved for review during future analyses. Indeed, such records might carry significant domain knowledge; both for training humans and computer learning algorithms, as well as testing proposed algorithms and human procedures.

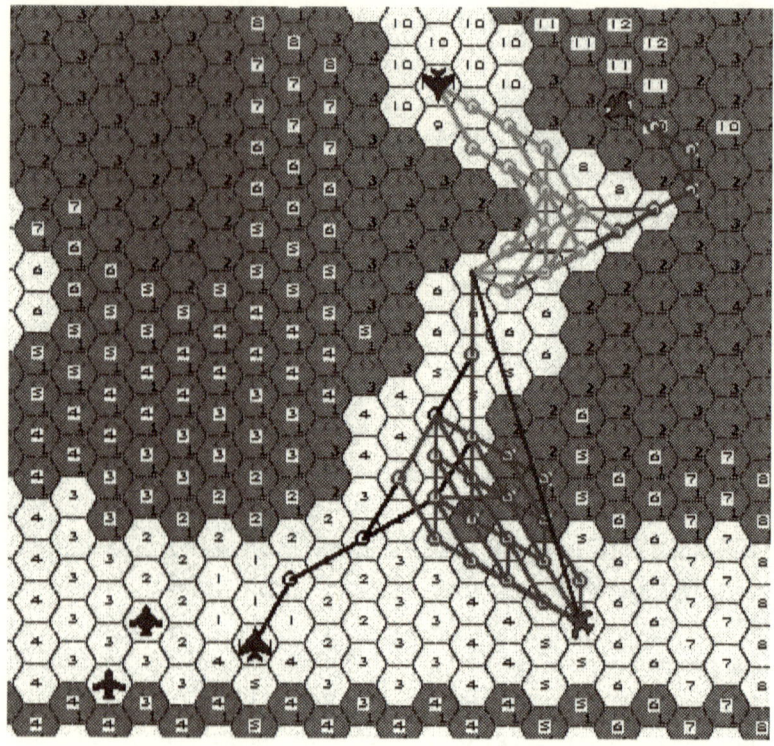

Figure 14-1. The figure represents an overhead view of a mission strategy, as generated by Mover.

MOVER AND SHAKER WORK TOGETHER

While the following scenario discusses UV operations, the LG/FSMwP technologies could be also applied to manned units or platforms. Here, reflecting current LG/FSMwP development, as well as ongoing experiments using groups of aircraft units as single-mission-level pieces, the game board partitions the conflict region into cells two miles across. Naturally, other cell sizes, platform types, and unit grouping could be supported by these technologies as well.

In this scenario, the operations center uses Mover and Shaker both as planning tools and as online tools, giving commands directly to UVs. Run on a portable computer system a safe distance from the main conflict areas, Mover and Shaker would be in a marine vessel operations center off the coast, or on an airborne C^2 platform. There, they interface with electronic target databases and logistical planning software, as well as a communications system that allows unfettered two-way communication with UVs, perhaps through a satellite or enhanced digital communication channel. UV-specific Shaker software, lower in the hierarchy, would be located on UVs themselves, though copies would run at the operations center to support the simulations.

It is important to note that the amount of autonomy given to UVs is decided on a mission-by-mission basis. Generally, because each UV carries only those portions of Shaker that pertain to its group, there is a highly distributed, highly autonomous form of control. As such, UVs do not need to request decisions from central control—unless, of course, a local event sequence has led to a high-level observable event. Yet because all UVs carry the same Shaker portions, there is not only a redundant use of resources, but also a possible synchronization problem between the multiple Shaker copies. Alternatively, the full Shaker hierarchical model may reside at an operations center, in which case all UV coordination is done there—meaning virtually no autonomy.

Primary Targets

In the current scenario, the joint force area commander is planning an UV strike on a primary target of high significance. As this plan is set in motion, Mover and Shaker link with recent intelligence sources, inputting initial enemy air and ground asset distribution. Using a link to appropriate information sources, details for friendly assets tasked to this mission are also uploaded. During this process, the commander receives Rules of Engagement (ROE) requiring that no air strikes be made to a particular geographic region because of the high civilian population. This ROE is automatically implemented as a restriction of available friendly moves in the Mover game board.

As part of their mission preparation, the commander and his staff use Mover and Shaker as what-if tools, entering objectives, an order of battle, and time constraints, then running simulations for potential outcomes. Indeed, they may run multiple simulations to select the best initial states—the best configuration of friendly assets. The staff might also test different ROEs, for example, an air policy in which aircraft involved in suppression of enemy air defense precede fighters, which in turn precede a bombing package; or perhaps different no-fly zones. In these simulations, Mover generates initial strategies for each simulation, with adversarial forces controlled either by Mover itself or a separate intelligent adversary simulation tool. The commander and his staff then compare the results of such simulations to find the strategy that yields the most favorable outcome. Clearly, via simulation the what-if analyses enhance the accuracy of estimated outcomes.

Other examples of possible what-if analysis results include:

— Simulating several sorties with differently sized attacking forces to determine minimal escorts to specific targets

— Ascertaining additional favorable mission parameters—asset allocation, package assembly points, and so on

— Judging the time-to-target for a time-critical mission to determine whether more dangerous routes should be taken

— Weighing winning/aborting conditions against a commander's objectives to maximize success rates while minimizing losses

— Testing alternative ROEs—for example, when to return fire if being tracked by enemy radar or fired upon

— Examining different UV commands and grouping configurations to focus on the same target

— Testing proposed ingress routes to avoid enemy interception

After simulations, the commander chooses a particular mission plan and gives the go-ahead for the mission to begin.

Checking Pulse

During the mission's execution, Mover and Shaker monitor the situation board, which indicates the movement of packages and aircraft via the movement of pieces on the game board. A visualization tool shows the game board status, allowing commanders and staff to monitor the engagement's progress. Mover and Shaker also give the commander suggestions and projections for the next few time steps' expected results.

Through observations of actual force movements, losses or gains, and changes in mobility, the Mover game board is continually updated, either automatically, or—when necessary—through a human operator. As such, the operations commander's staff might combine observed enemy strategies with Mover's ongoing simulations to improve actual strategy. Indeed, at any time during an engagement the command staff can run simulations, once again using Mover as an online what-if tool, updating target lists in response to certain losses, forecasting outcomes, and fine-tuning strategies. Since Shaker translates Mover's game board commands into coordinated actions within groups of UVs, such simulations might also include changes to those Shaker models subject to weapons modifications, and so on.

Thus, in a battle populated by UVs, Mover directs the UVs to proceed to specific geographical cells, while Shaker tells them how to react and coordinate within these cells. However, with manned units Mover and Shaker advise—not command—pilots or navigators. Indeed, there are already instances of such advisories in current fighter jets—ground-tracking radar, for instance, that issues pull-up signals.

The Chain of Command

The relative importance of events that occur in the battlespace is reflected within Shaker by how far up the chain of command, embodied in the hierarchical FSMwP model, the event is observed. For example, if an UV is attacked and evades, one of two things might happen depending on the design of the FSMwP. On the one hand, the event might be recognized as important and so percolate up the model hierarchy to influence the actions of other units in the group and potentially high enough to influence the Mover game board. On the other hand, the event could be resolved locally by Shaker—without having it make an appearance on Mover's game board. In every case, though, Shaker generates the most effective action—which also maintains maximum safety per Shaker's internal model. For example, when an UV is attacked and evades, Shaker might deal this with immediately with a response to counterattack. If the UV successfully destroyed the attacking enemy unit, however, this would appear on the Mover game board as a simple removal of an enemy piece.

In the context of a successful mission, one, or several, of the online what-if simulations eventually progresses to a state that satisfies the conditions for winning—with such conditions including any number of criteria, including the safe return of friendly units. At this point, the end is in sight. Mover and Shaker are effectively saying that a winning strategy has been found and that, unless an unexpected move occurs, the winning condition will be satisfied.

At any time, of course, a commander can override the simulation's next proposed strategic move, thereby altering the game board—much as an unexpected observation would. Naturally, any such alteration would effectively put Mover in a new initial state, and its algorithms would immediately start computing strategies from the new configuration of the game board and suggest a next move.

PARTITIONING THE REGION OF CONFLICT

Board games can exist in three as well as two dimensions. One example of a possible spatial partition—one, in fact, that has already been used in ongoing simulations—is a three-dimensional layered hexagonal battlespace partition, although the underlying theories for LG and FSMwP are independent of the how the region of conflict is partitioned.

Because Mover only observes movement at discrete times in such spatial partitions, Mover performs its computations periodically—for instance, every 30 seconds. Therefore, those discrete moves reflecting battlespace changes within the last 30-second time segment all appear to Mover as if they occur simultaneously.

In general, any board game inevitably includes four items:

— Game board definition and relationship with conflict regions—the location, orientation, and dimensions of the partition blocks
— Rules defining both the pieces' legal motions on the board as well as their interaction with other pieces
— Pieces' initial placement on the board
— Conditions determining game outcome—winning and losing

A battle is represented as an adversarial game that takes place between multiple players on the game board. Each player has its own goals and is able to exercise at least partial influence over the outcome of the game. To that end, a strategy is an algorithm that generates the set of moves to be made at the next update based on observed moves from the previous time slice. Given such a game board frame, the task of LG technology is to supply players with strategies to achieve their goals.

Pursuing such games further, the most general are totally concurrent ABGs, in which all—or some—game pieces can move simultaneously at action times [4]. In addition, a necessary aspect of LG technology is the pieces' reachability between update times (or, put another way, limitations or restrictions on pieces' movements) [5]. As an example, the reachability, in one time step, of a long-range missile is illustrated in Figure 14-2. This also indicates the enormous range of possible next steps in the game board formulation, and illustrates why search algorithms can be intractable over game board evolutions.

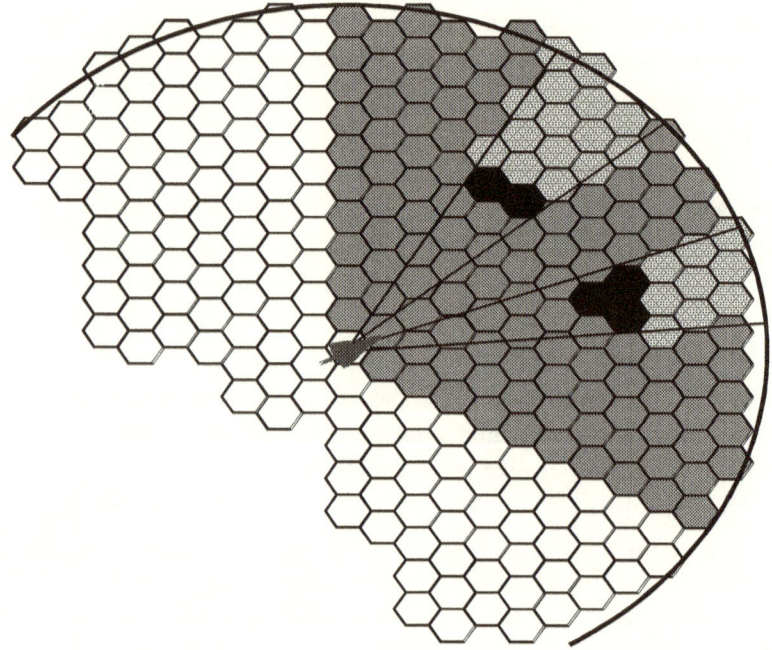

Figure 14-2. The shaded region shows the possible location of a long-range missile in the next time step. The darker cells show optimal lock-on regions.

CORDIAL EXCHANGE

Figure 14-3 presents a simplified data flowchart of the information exchange between Mover and Shaker. As indicated, the process starts with commanders entering operational specifications—objectives, for example, end-states, initial allocations, and so on—in Mover. (A human operator may need a tool to translate standard objectives into a form suitable for Mover.) Mover then acts synchronously: based on the set of events (c) received in the last T minutes (b), it supplies commands (d) to the pieces on the game board every T minutes. Mover also provides suggested courses of action, as well as alternatives, from which the commander may choose.

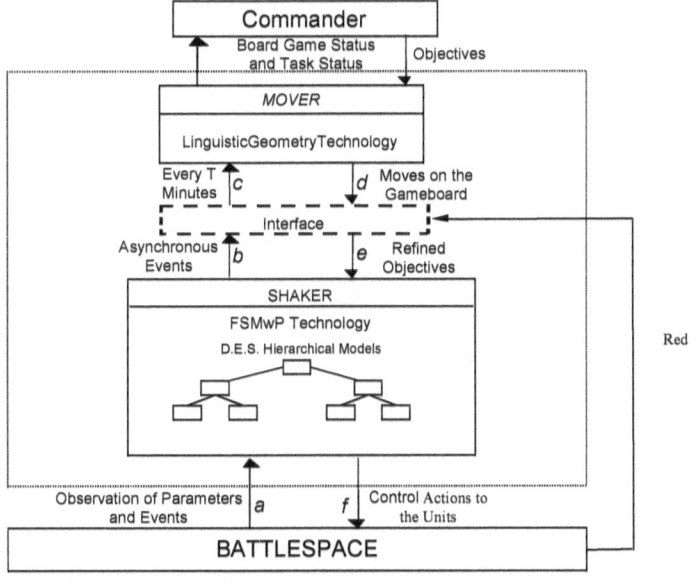

Figure 14-3. Information flows between a human commander, Mover, Shaker, and the battlespace.

Depending on actual events, interactions between Shaker and the battlespace (a and f) occur asynchronously. Figure 14-4 illustrates the relationship between the two types of feedback: in the data-flow diagram, Shaker's asynchronous feedback would be paths ab and f, while Mover provides synchronous feedback through the outer control loop path abc and def.

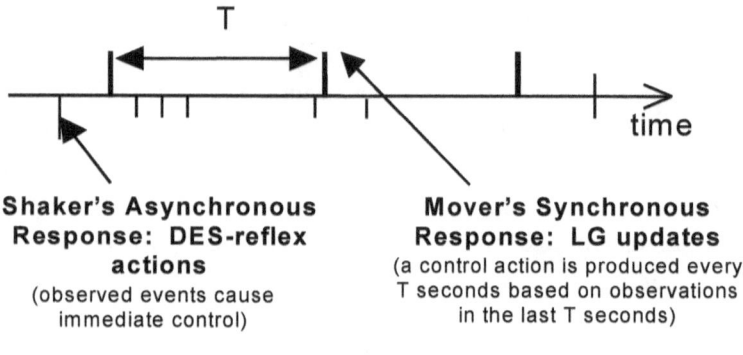

**Shaker's Asynchronous
Response: DES-reflex
actions**
(observed events cause
immediate control)

**Mover's Synchronous
Response: LG updates**
(a control action is produced every
T seconds based on observations
in the last T seconds)

Figure 14-4. Mover commands changes to the game board on a regular basis, while Shaker reacts at any time.

UNDER THE HOOD

Traditionally, finding strategies for ABGs required searches in giant game trees—that is, searching all possible futures. Instead, what LG does is dramatically reduce the size of the search tree by replacing much of the calculation with a construction based on formalizing search heuristics.

In practice, LG involves encoding small, localized, proven strategies (one-on-one responses to predictable enemy behaviors, for example, or tricks to control specific zones around a unit) as single symbols, much like the common linguistic habit of coining a phrase. Using the new symbols as building blocks, the LG algorithm then codes strategies at a meta-level—meaning that it constructs extensive strategies for large spaces in a very short time. The end result is a refutation of enemy moves—a refutation that will drive the board configuration (in other words, the location of the pieces) to states in which the winning condition (as defined by the commander in the operational specifications) is satisfied [6].

For its part, Shaker's key technology applies supervisory control [7] to models that include parameters that accompany discrete evolution [8]. C² system operators use the FSMwP framework as a design environment—in the same way that a flow chart is used in designing processes or software. Here, the nodes in the FSM graph are modes for the units, while conditions that trigger a transition label the arcs—with trigger conditions dependent on the model's continuous parameters. Models of individual components are stored in libraries for later use during the system-configuration that occurs in an operation's planning stage. In this phase, corresponding library models are given specific information—fuel levels, for example, weapon types and counts, and so on—to generate FSMwP models. Then, according to a force-composition or command/grouping structure hierarchy specified by the C² designers, and via abstraction and composition operations, models for higher-level aggregated entities (including the highest level, where the entities are pieces on the Mover game board) are generated automatically. First, based on detail models and corresponding mapping in the abstraction operation, models of an aggregated entity's constituent lower-level entities are generated. Then these abstracted models are synthesized into the model for the aggregated entity in an extended parallel composition [9].

Such a modular approach is extremely important, for it allows programmers (or, ideally, operators without programming skills) to combine individual models, thereby synthesizing complex FSMwP models for large groups of units. Based on this FSMwP model, then, when Mover sends Shaker specifications, Shaker performs online calculations, producing those control signals required to achieve Mover's specified actions. Furthermore, through the computations done over the higher-level aggregated entities—but enacted on the individual, lowest-level FSMwP models—Shaker not only issues specific commands to each unit, but also implicitly coordinates units. In addition, while these online calculations create safe and optimal policies for FSMwP models over the group of units, they also need to be distributed throughout the group. As such, controllers must project any newly formed policy into the individual unit. Their function will not cease, of course, for it is a key aspect of Mover and Shaker that, when intelligence information is updated, or mission objectives change, their twin process of re-computing policies from specifications repeats frequently during the course of a mission [10].

THE CHOICE OF GRANULARITY

On a larger scale, the approach could be used at higher levels of abstraction by employing multiple game boards at different levels of command as indicated in Figure 14-5. At the highest level, then, using large teams of mobile military units as game pieces, a strategic planner might employ low-resolution models to control global operations [11]. Clearly, LG reachability relations could easily be extended to larger groups of units—teams of submarines, for example, or ships, land-based military vehicles, airplanes, even space assault vehicles—and, as such, hierarchies of higher-resolution models could then be used to control smaller teams and focus on smaller operations.

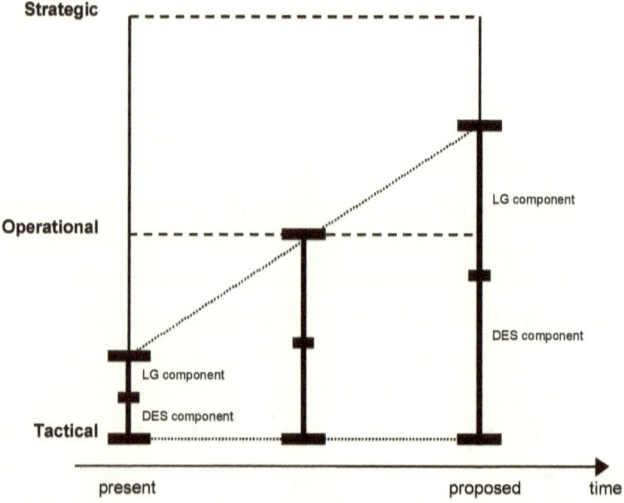

Figure 14-5. The current and planned abstraction levels demonstrate the application of the Mover and Shaker technologies.

This being said, performance may be sensitive to the choice of granularity. It is widely understood that coarse models lead to suboptimal solutions, while mismatches between model fidelity (in other words, how accurately a model matches reality) and requirement fidelity (how detailed the necessary actions/resources must be to achieve a goal) will lead to poor performance. For example, large physical partitions would prohibit exploiting a lack of enemy unit maneuverability. More specifically, larger hexagons would mean that aircraft would be able to transition to any contiguous hexagon; in the current implementation, aircraft can transition only to those hexagons directly ahead, to the left, or to the right. As such, there may be simple winning solutions to the board game, or no willing solutions at all—and therefore no benefit to using LG techniques over brute force search algorithms or simple inspection.

BUILDING THE BOARD GAME

During both planning and execution, LG and FSMwP technologies can support human decision making, as well as provide control signals directly to UVs. Hence, user interfaces are required at several levels to support interactions with a controller. First, a user interface must support the review, creation, and modification of the overall battlespace definition, which might in turn include grid size and location, as well as weapon and platform effectiveness, weapon system characteristics, and winning conditions. The interface would also have to support an automated extraction of much of this information from current intelligence data. Second, a user interface must allow a human decision-maker to view and, if necessary, modify the strategies produced by LG technology. Therefore, this interface would have to translate the LG output into a format readable by humans. Third, the user interface must allow viewing and modifying all FSMwP models. Again, the challenge is to translate FSMwP signals into a format that is both readable by, and has meaning for, the decision-maker.

Intelligence, surveillance, and reconnaissance data are also necessary to implement LG and FSMwP technologies. Clearly, the assumption is that information from sensors and reporting agencies has been collected and fused into a single input representing the current perceived truth about the battlespace. Currently, such data is collected at local facilities and reported up the chain of command on a regular schedule. To support LG time-scale requirements, there may need to be adjustments in the reporting periods; or, to allow the system to make asynchronous updates as required, there may need to be distributed access to resource data at the source.

AWARENESS AND ROBUSTNESS

A central concern about using an ABG approach to military planning and operations has been expressed by experienced military personnel: their common reaction is that it simply can't capture the situational awareness of multiple units in the heat of battle.

As stated in Chapter 2, the branching factor—that is, the number of possible next moves—and the implicit randomness of the engagements, mean that the transition structure for conflict can be very rich indeed. Therefore, the argument against capturing situational awareness with a board game usually hinges on the claim that such a rich structure for future events precludes computational planning. Yet LG strategy generation is different, for it captures expert localized strategies and polices as single atomic strategy symbols. LG's algorithm then searches through different combinations and sequences of atomic strategy symbols to form overall strategies, in effect reducing the search-space branching factor by selecting between small-scale strategies and single moves. In addition, the LG algorithm also avoids searching over useless nonsense moves.

A second concern with using the abstract game board approach is that of model uncertainty. In general, accurate models of a battlespace are difficult—if not impossible—to construct. There are, of course, numerous reasons for such problems, not the least of which are continuing changes in both enemy doctrine and technology. Thus, anyone concerned with battlespace modeling must be vitally concerned about whether a proposed technology will be robust—that is, will it be able to generate effective and accurate controls in spite of the inevitable errors in its internal models?

One reason to expect robustness from the proposed LG technology is that at each time step the algorithm recomputes future move sequences to counter adversarial strategies. Therefore, online control may be robust regarding some modeling errors in the same way that traditional model-predictive control schemes are self-correcting. For example, one cause of modeling errors could be an imperfect reachability relation; in other words, in reality, a piece can make a move that is not permitted by the game board's definition. One could envision how a chess player would have to recompute her strategy completely if her opponent were suddenly allowed to make an illegal move. When an illegal move is presented to Mover, by necessity it must reinitialize its configuration and begin generating strategies for this new configuration. Provided that illegal moves are not encountered too frequently, and are not dramatically different from legal ones, such a process can be expected to yield good performance in spite of any modeling errors provided the time horizon of the computed strategy is fairly long [12].

It is worth nothing that certain errors—enemy moves permitted in the game board but not possible in reality, for example, an enemy fighter with slower maximum velocity than expected—lead to conservative algorithmic control on the part of Mover. Yet such overestimation of the enemy also yields possible increased robustness, as Mover compensates for possibilities that don't exist. Of course, such overestimation—once again, of a possible enemy threat, causing Mover to react conservatively—also means it may not take full advantage of friendly force superiority. Although it might seem a conundrum, it is currently presumed that such conflicting modes will be offsetting, and a compromise will be reached.

In any event, as the level of abstraction increases it is natural to expect that the accuracy of the models—and the predictions made with them—will also increase. In general, aggregate measures are more reliable. Just as in betting $1,000 in a casino, the variance on a single $1,000 roulette bet is far higher than 4,000 $.25 bets at the slot machines. Though the expected results from the two processes might have approximately the same mean, the variance for the roulette wheel is much higher. Analogously, more reliable and accurate models may be available for groups of units than for single units. An air analogy would be that a flight package of 40 UVs applied to a target array might be predictable with relatively high confidence, while a single aircraft engagement might be very hard to foretell. Hence, modeling errors, and the corresponding need for robustness, will tend to be reduced at higher levels of abstraction.

The envisioned application also allows humans to override moves suggested by the LG algorithms. Such human intervention represents a similar kind of disturbance to that encountered in a model mismatch; if, for example, a piece were observed to make an illegal move. In both cases, an unexpected event occurs that effectively reinitializes the game board. Indeed, a description of the deviations from predicted outcomes that the model mismatch could produce—for example, a constraint that new positions must be accurate to one or two small changes of the board—is certainly of interest and a subject of further investigation. Advancing the research would obviously expand the possible use of Mover and Shaker to cases where enemy capabilities are poorly known, and where, similarly, there is significant uncertainty in the location and possible movements of the units in the battlespace.

SUMMARY

As in several preceding chapters, the focus is on a game formulation—specifically a board game—for representing the C^2 problem in a way that explicitly considers an intelligent adversary.

However, here Mover and Shaker decompose the C^2 problem hierarchically into higher-level generation of winning strategies on the game board, and lower-level, fast local reactions. The advantage of this division of labor is that each problem is much simpler than the combined one.

Mover is based on LG, a technique of intelligent reasoning in board-game problems that uses the formalization of search heuristics to build strategies—in effect, scripts that, in response to adversarial moves, specify how to move pieces. Unlike other approaches, as the board grows in size and complexity, LG remains computationally feasible. Mover periodically recalculates strategies, particularly when events invalidate previously computed strategies. Mover reasons at a higher level of abstraction, ignoring some low-level details—for example, a platform's specific mode. Mover's strategy becomes a specification issued to Shaker.

As events occur, Shaker uses the specification to generate and issue real-time control signals to physical units, or platforms—UVs, for example. Shaker uses a technology that combines finite-state machine with continuous parameters and thereby reduces the state space while still accounting for important considerations that tend to be continuous—for example, onboard fuel level. To build a model suitable to control a specific operation, Shaker automatically composes pre-defined models to create large models of groups of battlespace entities. Shaker's software could be distributed between a central command node and UVs' onboard computers.

Potentially, this approach can be used for dynamic replanning of an operation during execution and for issuing commands to UVs as well as suggestions to manned platforms.

CHAPTER 15

TEAM COORDINATION

Coordination within teams is commonplace in nature, for instance in schools of fish or prides of lions. It is also common in a variety of human-teaming endeavors, such as sports or music. However, in robotics, implementations are scarce in which autonomic robotic agents communicate and coordinate in order to achieve combined goals [1]. One of the main difficulties in coordinating groups of autonomic robotic agents is relating the detailed control routines encoded into each agent to the overall behavior of the group. A second difficulty arises when a single human operator attempts to coordinate the actions of a large set of autonomic units because the operator is unable to focus detailed attention at many distributed locations simultaneously.

One solution that overcomes these multiple difficulties would be to design the behavioral protocols for each unit in such a way that higher-level, goal-oriented commands can be given by a human operator to the group as a whole yet enacted by the units in a coordinated manner. Such a solution removes the need for an explicit link between low-level protocols and group behavior, and does not overload the human operator with fine control over individual units. The technology described in this chapter uses such an approach for the coordination of C^2 of military UVs. An analogy that serves well in discussing the proposed approach is the huddle, an integral part of American football.

A game of football consists of a sequence of plays, separated by breaks. During each break, and before the players reset into formation for the succeeding play, it is standard practice for both teams to huddle. In the huddle, each team's players gather in a tight group to be given the overall strategy, explicit and implicit roles, and actions for individual players during the next play. Several features of the huddle and ensuing play illustrate the proposed approach to automating the control and coordination of a team of UVs.

— Before each game, a playbook is created that identifies a finite set of fixed plays by name and lists individual actions for each player in each play. This playbook is based on an analysis of home team strengths and opposing team vulnerabilities. It is standard procedure for the players to memorize the playbook and recall it during the huddle.

— During the huddle, an appointed player selects a play from the playbook and communicates it to the other players. Individual players then translate the play into tasks that they perform.

— Players rely on their training to perform their respective tasks in a context-dependent manner. For example, when instructed to run deep into enemy territory to receive a pass, a player has some leeway to choose different routes depending on the layout of the opposing team.

— During the play, players coordinate among themselves in real time. For example, to prevent an opposing player from tackling a teammate, a player may block an opposing player.

LINING UP FOR THE HUDDLE

Researchers at the Pennsylvania State University Applied Research Lab, University Park, Pennsylvania, along with employees from Boeing Phantom Works, St. Louis, Missouri, are developing a technology, which applies distributed and hierarchical control concepts from DES to coordination in the C^2 of UVs. Such an approach—and more specifically, the collection of software routines located in each UV and manned vehicle, and in a central operations center—is known as the *UV Coordinator*.

In line with a football huddle, the UV Coordinator works in the following ways:

— Like the creation of a playbook, before the start of a military operation, officers responsible for preparing the mission generate a protocol for the UVs' behavior, then load the protocol onboard each UV. As in football, most protocols are generated by combining selections from a pre-defined library of low-level routines and would reflect special requirements or special knowledge about enemy capabilities. In the UV Coordinator approach, such low-level protocols take the form of FSMs that specify what actions should be taken by an UV when a particular event, or combination of events, occurs. Additional FSMs for higher-level protocols—that is, protocols for controlling the entire team of UVs—are also loaded into a computer at the operations center. In general, FSM design is based on information about enemy vulnerabilities as well as overall mission goals. For example, to reduce friendly losses operators might load a protocol in which damaged units must be escorted from the area of conflict. In toto, the collection of FSMs in the UVs and in the operations center is referred to as the Hierarchical Controller (HC).

— In the same way that names are used in the huddle to identify each play, high-level controllers at the operations center can give short sequences of commands, which are decomposed automatically via an UV Coordinator software into individual UV actions. As opposed to the huddle, where high-level control can be done only once per play, the UV Coordinator allows high-level commands to be issued at any point in time.

— Officers responsible for mission preparation also produce a collection of context-dependent action generators, known as the Tactical Intelligence System (TIS), and then load these generators into the UVs. As opposed to HC, which gives discrete commands from observed events, these generators translate discrete commands—attack target, say—into continuous control routines for the UVs. For example, during the mission, a specific flight path and best on-board munitions might be selected to counter specific enemy defenses. As with HC, TIS control routines are usually selected from a pre-defined library. Here, this selection can be based on standard military doctrine and an analysis of enemy vulnerabilities, or on simulations that test likely outcomes for different TIS routines and optimize projected performance.

— In the same way football players must coordinate real time on the field, some UV coordination will take place without any need for higher-level human intervention. For example, once an officer introduces an UV protocol in HC indicating that damaged UVs must be escorted; UVs can coordinate locally to ensure that this action does in fact occur. This coordination is accomplished through the sharing of observed events between UVs. Depending on the importance of the observed event, it is relayed to higher-level components within HC and coordinated actions are sent back to the UVs whose planned maneuvers are impacted by this observation.

In the UV Coordinator approach, the HC makes decisions based on pre-loaded protocols in the form of FSMs as well as inputs from the environment. Then, based on the current situation, TIS works out the details of how to implement these decisions.

The main advantage of the UV Coordinator approach is the ability to command multiple vehicles with single or short sequences of commands. Such an approach reduces a human supervisor's workload, enabling him to focus on more important issues and decisions, such as how the group should behave as a whole.

The use of HC brings with it two specific advantages. First, because of the approach's rigorous theoretical foundation, is the ability to verify, in a purely objective manner, that specifications for the UV protocols—for example, the absence of deadlock (see Chapter 3), or each UV's ability to abort independently—are actually satisfied. Second, depending on scope and importance, HC supports decision making at different levels. For instance, the on-board FSM might immediately handle a reaction to an onboard component failure within a single UV, while an event corresponding to the destruction of an UV might percolate sufficiently high up the chain of FSMs to be flagged for notice by the human operations commander.

Finally, the modularity of HC and TIS offers a method to combine generic or doctrinal strategy, which might appear in the libraries of HC and TIS routines, with conflict-dependent or weapons-dependent strategies, which would be generated during the lead-up to the actual operation.

COORDINATOR IN ACTION

The following scenario illustrates the use of the UV Coordinator in a hybrid manned-unmanned air operations strike mission. The scenario does not imply that this technology should be restricted to air operations. While here the UVs are under the control of fighter pilots, in other applications the UVs could just as easily be under the control of a ground station, ship, or airborne command. Similarly, the UVs may be performing a variety of other missions—autonomic reconnaissance, for example, or locating and destroying a column of armored combat vehicles [2].

Here, the players are a flight of four fighters, with an attached escort of five UVs tasked for an interdiction mission in a major regional conflict. The fighters, with precision radar-warning receivers, have a broadband datalink between their cockpits and the UV control units. The UVs, small, low-observable, subsonic unmanned aircraft, are capable of carrying guided munitions over greater distances than the fighters. Equipped with the datalink to receive controls issued by the pilots, and a simple radar transponder, the UVs also have radar capable of moving-target indicator functions to 20 miles. (Here, primitive automatic target recognition is assumed.)

In both the fighters and UVs, the UV Coordinator software forms part of the operational software. In the fighters, the software is integrated into the voice-warning system as a decision aid to provide critical advisories to the pilot as well as to permit commands to and status-checks from the UVs. In the UVs, the onboard UV coordinator software is part of aircraft flight controls.

The diagram, Figure 15-1, indicates information flow, with the letters used to identify specific information channels.

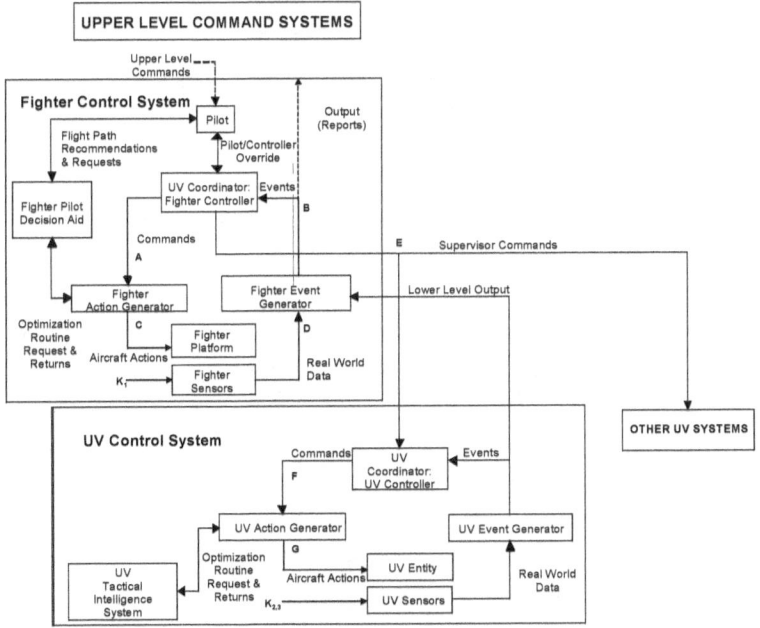

Figure 15-1. Information flows between pilots and UVs during the mission.

The mission is tasked to conduct a precision attack against 8 aimpoints at an Intercept Operations Center supporting the enemy air defense network.

The flight formation, illustrated in Figure 15-2, shows datalink pathways (dashed lines) that can pass both commands and data. Pilots and airborne command elements are the humans in the loop, or in this case, the hierarchical control loop. Commands are simple: a pilot can give an UV commands, including escort, attack, follow, decoy, return, defend, and relay. In turn, the UV Coordinator provides pilots UV information in the form of aggregate—or high level—observations, including command acknowledged, enemy detected, damaged, low on fuel, target hit, and so on.

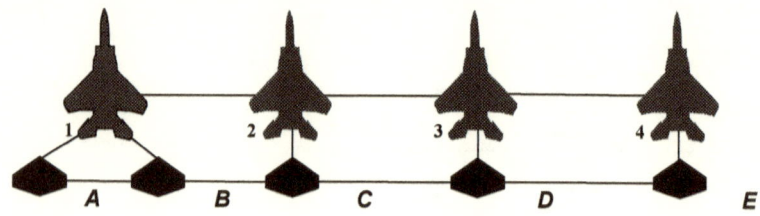

Figure 15-2. The pilots of Fighters (1-4) command the individual UVs (A-E).

Each pilot-generated command corresponds to a set of possible UV actions. For example, each UV may have multiple escort algorithms: once the upper level (here, the pilot) gives the escort command, the UV can only perform actions necessary to complete this instruction. Further, if the upper level hasn't specified a particular algorithm, then the UV is free to escort in the way its TIS deems appropriate. Here, it's up to the TIS action generator to translate the control system's discrete language into the rich movements of an aircraft. In this case, since the default command is escort, the HC is initialized so that all UVs are performing the escort command. Specifically, escort here instructs an UV to act as Suppression of Enemy Air Defenses (SEAD) escort, automatically attacking any radar or surface-to-air missile (SAM) systems that threaten the flight.

During mission planning, specific mission information and instructions are loaded into the UV Coordinator, including communications plans, crypto codes, terrain elevation data, weather data, and current threat. Used by the UV TIS as well as by the fighter's pilot decision-aid software, this information includes assignments enabling the UVs to join their host aircraft. In addition, mission information and instructions—including a rendezvous point and time—can be transmitted electronically to another base, and then loaded into UVs also flying this mission.

As expected, the aircraft launch on time, are air-refueled, and then meet up with the five UVs. The fighters deploy into formation, with elements split by eight miles as in Figure 15-3, and proceed at medium altitude to the border.

Figure 15-3. The fighters and UVs form an ingress arrangement.

After the fighters cross the border, the enemy scrambles two interceptors. Airborne command elements detect the launch and put the information on the link ($K_{1,2,3}$). At 50 miles, the lead fighters detect the enemy (K_1) and declare them hostile. At 30 miles, the fighters target the interceptors; fighters #1 and #2 lock on to a single interceptor, then command their UVs to go to Decoy (E). The UV Coordinator component accepts this upper-level command, then initiates the corresponding lower-level actions. As part of their action, the UVs turn on their radar transponders, thereby increasing their radar signature, and then maneuver, confusing the enemy. As the UVs maneuver, the UV TIS routines keep them from interfering with each other.

TIS, in the pilot's decision-aid, notes the enemy lock-on, determines the worst-case enemy missile range, and checks the closure rate. At 25 miles, fighter #1 shoots, but due to enemy interceptor maneuvering and jamming, fighter #2 cannot.

At 20 miles, enemy interceptors lock on. Interceptor #1, confused, locks onto UV B, while the second interceptor successfully locks onto fighter #2 (see Figure 15-4). TIS, in fighter #2, notes the radar lock and advises the crew (K1 then C). At 16 miles, fighter #1's shot explodes the lead interceptor. While fighter #2 is still trying to acquire the target, TIS calculates that the other interceptor is within fighter #2's weapons range and has a radar lock. The crew, still trying for a radar pickup, is unaware of the threat. TIS sounds a warning system to abort (C), and while both lead aircraft turn 180 degrees out from the threat, the UVs continue their decoy tactics.

Figure 15-4. The enemy intercepts the lead group, and the pilots command the UVs to go to *decoy.*

The trailing element now picks up the threat, and fighter #3 locks on, while fighter #4 continues to search for other threats. Interceptor #2, realizing that the target has aborted, searches for another target and locks on UV C (see Figure 15-5). UV C's radar warning receiver detects the radar lock-on (K_2). The UV Coordinator issues a command to evade the threat (F), and

TIS subsequently turns off the transponder and maneuvers (G). When the lock is lost (K_2), the UV controller again changes state and commands to a normal, stealthy escort mode (G). The interceptor pilot, still searching for a target, runs head-on into a missile from fighter #3.

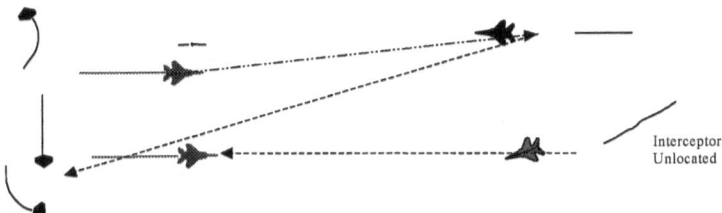

Figure 15-5. One interceptor fires on an UV, and the fighters lose radar lock on the second interceptor.

Due to the abort, fighters #1 and #2 are now heading in the opposite direction. Realizing this fact, the fighters turn around to form up and trail eight miles behind fighters #3 and #4. UVs A and B are commanded to escort once again (E).

The enemy air defense commander, noting the loss of his two fighters, clears a SAM system to engage the flight as it approaches the target. A SAM radar engages fighter #4 from 25 miles, but the fighter maneuvers to defeat the enemy shot. Fighter #3 follows. UVs D and E receive information that their host aircraft are defending (K_1, K_2), and the UV Coordinator, recognizing target proximity, issues a command for UVs C and D to attack (F).

For each UV, TIS translates this attack command into detailed trajectories and action sequences, and then commands the UVs to commence their attack along different axes. To facilitate the attack further, the TIS action generator coordinates the UVs' behavior to avoid duplication of effort. At 20 miles, the UVs use their radar to map the suspected radar location, detecting five radar images quickly identified as the radar and four unspecified vehicles. Still 20 miles out, the radar moving-target indicator on UV D notes that one of the target vehicles is moving (K_2). So alerted, the UV Coordinator response is to attack the moving target with higher priority (as prescribed by HC uploaded at tasking), and with two weapons instead of one (as prescribed by TIS uploaded at tasking). These responses are passed to UV E, which reprioritizes its plan, now moving to attack the radar and three remaining vehicles with one weapon each. The UVs attack the SAM system, destroying the radar, moving vehicle, and tow stationary vehicles, but missing the remaining vehicle. The UVs, making a follow-on pass, determine that the radar signatures have changed for four targets, but not the fifth (K_2). As a result, UV D reattacks (F) with a single weapon, destroying the target. The attack run on the SAM site is now complete: UV E has no weapons remaining, and UV D has one.

Prior to the UVs' attack, the SAM site launched one missile that catastrophically damaged fighter #4. The crew ejected, while the remaining three flight members continued to defend until the UV's first attack pass. Upon the loss of fighter #4's data link signal, UV E's controller automatically switched to come under the command of fighter #3 (a state change in the FSM of the UV Coordinator), but continued to execute the last command, to attack the SAM site.

Upon completion of the attack, fighter #3 commands UV E to relay (E). The UV's TIS determines the approximate location of the downed aviators, indexes the database with terrain and known threats, and chooses an orbit to act as a radio relay for the downed crew. To avoid highlighting the evading crew, UV E offsets 20 miles from the crash site but stays within radio line-of-sight. This position allows crew members to use a low-power function on their survival radios, thereby reducing the chance of enemy detection. The remaining three aircraft and four UVs proceed to the target area.

As the flight nears the target, UVs A through D are commanded to decoy (E). To confuse any remaining defenses, TISs on the tasked UVs command the UVs to augment their signatures and make a mock attack run on the target. Due to its previous target attacks, UV D is lower on fuel and therefore cannot both complete the decoy maneuver and return to base. HC dictates return to base (F), and TIS both plans the route and advises fighter #3, which in turn does not override HC's orders but instead lets the UV return to base on its own (E).

The fighters make a single successful pass against intercept operations, then egress at high speed. UVs A through C recognize the fighters' egress, and then linger in decoy mode for a short time at the target. When the UV Coordinator issues a return command (F), TIS plans a route for the UVs' egress.

UV E, acting as a radio relay, remains on station until forced to return to base for fuel. By that time, there is another pair of UVs on station, launched under a combat-search-and-rescue tasking, capable of attacking any enemy ground forces that might pose a threat to the downed crew.

The fighters, with their UV escort, have heavily damaged the target, destroyed a SAM battery near the target, and killed two enemy interceptors—all at the cost of one fighter. A search and rescue mission remains ongoing.

The benefits of the UV Coordinator to the mission are clear. By instructing the UVs to act as decoys, and to attack the SAM site while the pilots focused on the target, pilots were able to increase their own survivability and improve the probability of mission success. In addition, by having the UV Coordinator translate pilot commands into specific UV actions, the pilots were able to focus on overall mission goals. Pre-programmed responses in the UV Coordinator also enabled an UV to act as communication relay to a downed crew.

A PEEK BEHIND THE SCENES

The core component in the development of this technology is the Discrete Event Controller (DEC) test bed, which encompasses all necessary components for modeling the real world environment, abstracting events from that environment, and constructing commands for that environment. Figure 15-6 depicts the structure of the development system for one entity in a distributed architecture of coordinated agents (UVs, for example). Each entity uses such a structure in making decisions and performing actions.

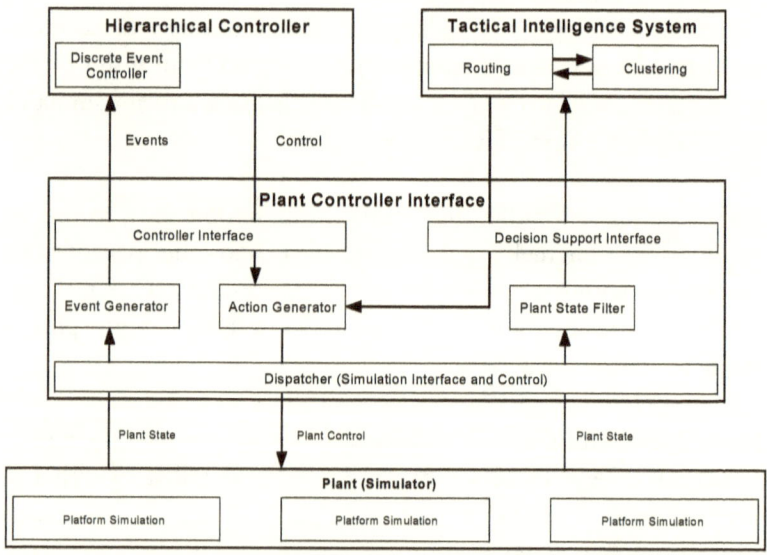

Figure 15-6. The development testbed shows the internal components of UV Coordinator.

The plant simulator is an Air Operations simulator to test TIS and HC algorithms. The sole interface to this system is via the Dispatcher, which uses a Common Object Request Broker Architecture bridge defined by an Interface Definition Language document. This document defines the key commands used for accessing information from the Air Operations simulation worldview and for issuing commands to the system. In the prototype development system, typical examples of these commands are move, fire, and loiter. Available information includes Blue force platform status and sensor data that tracks **Red** force movements and actions. Noise can be introduced into sensor information at the plant level to simulate the effect of bad sensors or poor weather conditions.

The event generator and action generator are the primary interface between the continuous world of the plant and the discrete world of the controller. Once every two seconds (an arbitrary time limit), the event generator reads in state data from the plant model. Then, using a specified set of rules, the event generator constructs a list of events. In terms of mathematics, the event generator is a parameterized mapping of real-world state data onto a set of abstract events. For example, in the current implementation, an alarm event is generated whenever a plane comes within 10 kilometers of a target. Therefore, a given plane's event generator will observe a plane's location with respect to known targets. Once the plane breaks the 10-mile boundary, the aircraft's event generator will signal an alarm, similar to the scenario's warnings alerting the pilot to the presence of enemy radar.

For its part, the action generator receives commands from the controller and translates them from the abstract language of events to the real world of motion and action. The action generator is therefore a mapping of events to a set of real-world actions. One event command—attack, for example—need not map to one real-world action. In general, event commands are translated into a highly complex series of real-world actions, consisting of moves, fires, loiters, and so on. The recipe for this sequence comes from TIS, which encompasses routing, firing, escape path, fuel usage, target selection, and weaponeering algorithms [3].

The hierarchical controller is a set of FSMs arranged hierarchically, detailing how—based on observed events—an entity should behave.

Here, each FSM acts as a controller, seeking to guide the entities it controls—that is, FSMs lower in the hierarchy—to one of a pre-identified set of states. In that FSMs receive symbols generated by the event generator as input, the FSMs are event recognizers. Therefore, once a symbol (that is, an event) is received, the FSM makes a transition—it changes state in the finite state-machine graph, then outputs a new control symbol to the entity it controls.

The theoretical foundation to the hierarchical controller has its roots in lattice theory [4]. Over the last 20 years, the field of DES, and the framework of supervisory control, has developed as the result of the application of lattice theory to systems modeled by finite automata [5]. In the supervisory control framework, the control symbols are actually inhibitions of future events rather than a single forced event, while the set of disablements is actually computed with a maximal controllable sub language computation [6]. This process of observing events, then reacting by disabling future events, embodies the decision-making process of each FSM in the hierarchical controller.

Each FSM interacts with other FSMs above it and below it in a hierarchical chain of command. When a lower-level FSM receives an event—an alarm, for example—the FSM may propagate that event up (still in the form of an event symbol), even though not all information needs to be passed up all the time. Deciding which information is important is a design consideration. The FSM at the next level (a supervisor) receives this event, and this reception triggers a transition in the FSM's own state structure. Once again, the FSM may possibly propagate the event to a higher level. Ultimately, a final level is reached, where a new command event is generated.

In the extreme case of a critical event, the top level reached might be a human operator outside the HC itself, who will return a command. In either case, the new command event is passed back down through the hierarchy, until it reaches the original controller that received the event. The new command event, then, enables a set of transitions at the lowest level in order to promote desired behavior. Here, the command event is also passed to other controllers in the hierarchy at the same level as the originator of the event. This mechanism of percolating the events up the hierarchy to a top decision-maker, then broadcasting the decision to all subordinate FSMs, is how the hierarchical controller coordinates the lower-level FSMs—and hence the UVs that the FSMs in turn control [7].

The TIS is composed of a set of algorithms that use real-world state data to compute best path routes, cluster targets, weapon choice, and so on. The action generator then uses these algorithms when it translates discrete commands into continuous control for the UVs. Based on current UV requirements, the action generator will select the best algorithm: one routing algorithm might be suitable for minimizing fuel consumption, while another might minimize exposure to threats, and so on.

Throughout, the UV Coordinator combines a discrete control device (HC) with context-dependent selection of the continuous algorithms (TIS). A major benefit of this approach is that an operator at any level of the HC can inject commands. In turn, these commands events are interpreted and acted upon by the lower levels of the HC and TIS algorithms, resulting in automatic coordination of the entities controlled.

GAPS AND BARRIERS

A possible limitation in the DES technology is that in the context of a complex, large-scale military operation, the construction of FSM proves too time-consuming—and is therefore potentially infeasible. Often referred to as the model-building problem (discussed in Chapters 2 and 3), there are two plausible mechanisms for reducing design time for those FSMs necessary to be loaded into the UV Coordinator. The first is the use of modularity; that is, FSMs are created during peacetime for individual platforms, or perhaps for specific behaviors of individual platforms, and subsequently combined with a parallel composition immediately prior to an operation or mission [8]. To a great extent, the UV Controller uses modularity. In the presence of high levels of heterogeneity in platforms and their behaviors, and because the generation process may still be overly time consuming, modularity becomes less effective. It is also likely that UV functionality will increase in scope and complexity in the future. Therefore, a second method—that may remove the design task completely—is the possibility of future tools that will learn FSM structure from prior engagement data or human behavioral tests

Robustness is also a fundamental difficulty in using DES to control ostensibly continuous devices. The process of matching signatures to events—that is, detection of events—can fail and lead to de-synchronization and erroneous control commands. Designing for robustness—that is, good performance even in the presence of modeling errors—is ongoing DES research, most notably with approaches based on language subset inclusion and single errors in the automata structure [9].

The most demanding requirement for UV Coordinator implementation is the development of user tools for FSM and TIS algorithm implementation and real-time UV control. Here, a database of FSM and TIS algorithms for individual platform types must be developed and maintained—a database that would necessarily be linked to a simulation environment incorporating a model of the battlespace. Different TIS routines and FSM designs for the hierarchical controller could then be tested—even immediately prior to a mission. For example, several attack profiles could be developed and tested to evaluate an individual platform's responses to specific threat scenarios.

Operator interfaces are required to allow human operators to verify platform configurations and their on-board controllers. Operators will also initially load those algorithms that reflect the rules the UV controller will apply. During execution, and in order to override UV actions when necessary, a human interface will also be required to allow a human operator to monitor individual platform readings and then interpret these readings.

Reliable communications channels linking UVs and supervisors are a necessary precursor to implementation. Because, for the most part, only command strings will be transmitted, bandwidth on these channels will not be particularly high. However, to interpret the sensor data and translate them into FSM symbols, pre-processing is required onboard the UVs. Indeed, pre-processing implies an extensive database of typical signals representing known threats and targets. For example, an individual platform may receive a heat signature from a potential ground-based target. At the same time, there may also be a target image from optical scanners. These two distinct pieces of information must be processed, then compared to typical signatures from known entities; finally, a decision can be made as to the nature of the sensed object. This procedure will probably be done on a country-by-country or regional basis, so to eliminate as many low-probability choices as possible and to speed processing time. A set of rules will also be needed, to allow the processor to deal with incomplete and potentially conflicting inputs. These rules may also include the requirement for human intervention in cases where no clear identification or conflicting identifications are made.

SUMMARY

It remains a technical challenge to construct algorithms that will make a group of robots—UVs, for example—act as a cooperative, coordinated team. The approach presented here—Hierarchical FSM—uses distributed and hierarchical control concepts from DES, and can be compared to a huddle in football. In football, a team has a pre-designed playbook, the leader selects a play, the

players execute their tasks, and as events unfold coordination occurs
in real time.

In the overall system, the Hierarchical Controller module is a
collection of protocols that, when events occur, specifies what an
UV should do. From a library of lower-level routines, and for each
specific mission and threat, protocols are synthesized semi-
automatically (with human guidance); protocols then take the form
of FSMs distributed between UVs and the remote operations
center. The automated synthesis overcomes the notorious model-
building problem, and the correctness of the synthesized
protocols—potentially large and complex—can be automatically
verified in a formal, rigorous manner. Similarly, mission operators
guide the synthesis of the Tactical Intelligence System—a collection
of routines for computing detailed actions and paths—for
uploading to UVs.

From the operations center, a single human operator, through
simple short commands, remotely controls a large, complex team
of UVs with heterogeneous capabilities and roles. Onboard the
UVs, HC/TIS software translates high-level commands into specific
detailed actions. With controls produced by HC/TIS, UVs
coordinate their actions locally, without the need for further
operations-center intervention.

PART VI
CONCLUSIONS

CHAPTER 16

RETROSPECTIVE:
AUTOMATION AND HUMANS IN COMMAND

The technology concepts presented in this book imply a radical invasion in what was heretofore an exclusive domain of human decision-making—military command, a most hallowed ground of human intellectual endeavor in which the life and death of men and nations was decided. Naturally, such concepts must meet a number of profound concerns, none as important as the challenge of relations between humans and computers that automate aspects of a human's decision-making process.

In each of the preceding chapters, we paid attention to a few issues of human factors, or human-computer interfaces, specific to the technology at hand. Those discussions, however, merely scratched the surface. In this chapter, we look back at the implications of the proposed concepts, attempting a slightly broader and deeper discussion of difficult, overarching issues related to the incorporation of new automation in C^2. It is this impact, ultimately, that will decide the value and fate of this book's ideas.

Before proceeding, we must clarify what we mean by the term automation. For the purposes of the following discussion, we consider as automation any process in which computers combine or transform information without direct human involvement. Even a word processor, therefore, would be a form of automation. In line with this broader definition, concepts of operation for automation vary according to underlying variations in the degree of authority and autonomy delegated to computers. At one extreme are fully automated controllers that have the authority and autonomy to make decisions and take action without human intervention, at least in principle. At the other extreme are systems people generally call automated decision aids or decision support systems. These systems may not act on their proposed decisions but are there to provide information to humans who have the authority to act. Most systems, including the types of systems described in this book, fall somewhere between these two extremes.

Although a variety of automation technologies exist, they serve three basic purposes. Automation performs tasks that humans can't do, tasks that humans do poorly, or tasks that humans find undesirable [1]. Automation makes it possible to fly stealth aircraft despite their complex aerodynamics. Automated cameras improve on people's ability to convert quantitative estimates of light into the shutter speed necessary to expose the film appropriately. Automatic automobile transmissions remove the tedium of shifting gears [2].

The automated tools discussed in this book support decision-making tasks, which in situations characterized by large amounts of information, uncertain outcomes, and time pressure can challenge human decision-makers. Indeed, the technologies described herein address aspects of military decision making that just 11 years ago were considered unsuited fora automation [3]; that is, these tools offer support for the higher-order cognitive processes typically associated with operational and tactical planning and execution. Task Manager, for example, as described in Chapter 8, generates options for strike-force composition that control long-term risk to task success in terms of task deadline and friendly-force losses. Auctioneer, as described in Chapter 10, generates options for allocating tasks to strike and support units by considering the risk associated with planned routes as well as the synergistic effects of multiple missions in the bigger operational picture.

By automating aspects of military decision-making, such technologies can potentially counteract human information-processing foibles that have been consistently demonstrated by more than 50 years of basic and applied decision-making research [4]. These foibles stem from ubiquitous limitations in human information processing capacity: people have a limited amount of attention, and hence can process only so much information in a given amount of time. This attention limitation causes people to take predictable short cuts, while making decisions under uncertainty. Time-pressure serves only to increase the use of these short cuts (that is, heuristics).

The effect of using these heuristics is that complex decisions are often made without considering all relevant information. Selective information processing, in turn, often leads to the implementation of suboptimal options. By contrast, the tools proposed in this book select options based on a thorough analysis of information deemed relevant. Thus, an underlying theme in this book is that the proposed algorithmic approaches to action selection can be used to augment the decision-making heuristics employed by humans, thereby enhancing military operations.

This book, therefore, conveys a vision of a cooperative decision-making situation, where information and decision-making authority are distributed among humans and automation. This distribution means that humans and automation will have to work together— if they are to achieve the coordination required for concerted effort toward common goals. Consequently, if these technologies are to be useful, they must be good team players—in addition to being good at what they do.

This chapter describes aspects of the interaction between humans and automation in complex systems. The general impact of automation on humans is described first. The discussion then turns to four points of particular relevance to the technologies in this book.

THE WINDOW ON THE WORLD

Viewing automation as a window on the world of work highlights automation's impact on humans in the system. In this view, automation serves as an interface between humans and the work they do. The work described in this book is planning and execution, with technology being inserted between humans and the job of planning, monitoring, and controlling.

The most salient change produced by the introduction of automation is to distance humans from the work they do. Instead of doing the work, humans supervise work as it gets done, providing direction and monitoring progress. The automation actually does the work (for example, option generation and analysis) and reports its status (for example, makes recommendations as to which option should be implemented).

Interaction requirements created by this supervisory arrangement can be very subtle. As a simple example of the subtleties this arrangement implies, consider a fly-by-wire system in a high-performance aircraft. When F-16 pilots first tried fly-by-wire, they had trouble determining airspeed and maintaining proper flight control [5]. Given that the necessary information was readily available on cockpit displays, the pilots' troubles were surprising. As it turned out, the root of the problem was neither automation nor pilot skill. Instead, it was the interaction. Pilots relied on vibrations in the flight-stick, but the stick no longer vibrated. Now, artificial stick-shakers are routinely added, providing pilots with the proprioceptive feedback they expect. The technologies proposed in this book may present an even more challenging design problem.

The reality of the supervisory relationship between human and automation becomes more apparent if you imagine being a pilot during an automated carrier-landing. You should watch intently, so that you could take over should things go awry. You should constantly compare your automated copilot's performance with your knowledge of how a safe landing should progress. And you'd better be ready to grab the controls.

The imagined pilot's response is generally what humans do when they interact with automation. They monitor it to ensure prudent operation, and stand ready to take full control if it malfunctions. Indeed, the decision of whether automation is working correctly is fundamental in human-automation interactions, and one of the main reasons humans remain an integral part of complex systems. Automation excels at making normal operations more efficient, but humans are better at adapting to aberrant or idiosyncratic operating conditions; consequently, humans can provide a stabilizing influence on system performance.

As part of this arrangement, humans face attention-management issues and additional knowledge requirements. Humans must divide their attention between automation performance and the work they are trying to accomplish; doing so effectively requires knowing how to accomplish the work and how automation may do it differently. In the context of this book, such an understanding means, for example, knowing what a reasonable action is and why the automation's recommended (or enacted) action may differ.

Should it appear that the work is not being accomplished, the supervisory arrangement also requires additional diagnostic and troubleshooting skills. When automation is involved, what was once one decision becomes three [6]. The automobile dashboard light that tells you when you're low on oil provides a very simple example. Anyone who's had an old car knows to check the oil before driving it. If the oil is low, you add some, and then go. In newer cars, this system status is monitored automatically—simple automation, but automation nonetheless. If the light comes on while you're driving, maybe you need to stop and add oil. Or maybe the light is malfunctioning. It could also be that the sensor should be replaced or recalibrated, so that it will quit issuing false alarms.

Paradoxically, automation can add workload at points where decisive action may be required to avert catastrophe. When operations appear to be going sour, operators of complex systems, unlike drivers, often do not have the luxury of stopping the car to check the oil. Unlike pilots, they often do not have the option of prying their eyes off the instrument panel and looking out the window— because the instruments are the only window into ongoing operations. If the window-on-the-world indicates that work is not progressing as expected, the human faces a fundamental decision as to whether there is indeed an unexpected or unsafe work condition. Diagnostic decisions about whether the automated system is functioning as intended, and whether the indicators in the interface between human and automation are functioning properly, feed as evidence into this fundamental decision.

In terms of the technologies described in this book, the complexity of the decision-making situation, as well as a lack of time, may prevent operators from looking out the window. By the time the human realizes that a recommended action is inappropriate, manual replanning may not be an option. Thus, operators may well be put into a position where they have to determine quickly those reasons underlying faulty recommendations in order to get the technology to generate a reasonable action. While this operation may involve adjusting parameters used by the automation, it could also involve fixing a relatively simple technical glitch—such as plugging in an information feed that's somehow been unplugged. The technologies described in this book must support these troubleshooting tasks by allowing operators to look under the hood when necessary.

Thus, automation changes jobs by creating new tasks, knowledge requirements, and possibilities for error. Monitoring automation often requires a sustained investment of attentional resources. Distancing humans from work requires that humans have different types of knowledge about their tasks so that they can effectively supervise operations. Here, humans must internalize models of how the automation works—at a functional level—so that they can predict how it will react to different inputs. In addition, human supervisors will have to troubleshoot apparent discrepancies between expected and current system behavior. One consequence of these new automation-related tasks and knowledge requirements is the possibility for new types of human error, or, depending on one's perspective, latent system failure [7].

FOUR POINTS OF IMPACT

If the technologies proposed in this book are developed into fielded systems, they will create new human roles that are supervisory in nature, and corresponding changes in attentional focus and knowledge requirements. The new supervisory roles will effectively shift the level of abstraction at which some military professionals work. For these personnel, a more abstract level of reasoning will be necessary, as they interact with the new forms of corporate knowledge embodied in these technologies to accomplish their task.

As described below, these new roles, and new forms of corporate knowledge, can potentially alleviate some of the decision-making difficulties experienced by humans. However, before these potential benefits can be reaped, methods for productive interactions between humans and the new automation must be developed. The biggest challenge involves creating ways to communicate intent accurately to these technologies so that they produce relevant recommendations or controls. Representational aids and explanation facilities are also needed in order to foster the interactions between humans and automation.

New Supervisory Roles

For some personnel, automation will tend to shift their attentional focus to a higher level of abstraction. In particular, the roles of those personnel currently involved in the detailed generation and analysis of possible actions to achieve their commanders' goals will change such that they take on a flavor of supervision. These personnel will no longer be concerned with the details of generating and analyzing possible actions because the automation will handle the details. Instead, they will concern themselves with defining goals and objectives in a manner that conveys their commanders' intents to the automated tools they use. Their new supervisory roles will also require them to adjust sets of interrelated parameters as they attempt to fine-tune the recommendations or controls their automated tools generate. Thus, tomorrow's operators will find themselves in a situation where they direct automated tools, guide them toward reasonable actions, and assess the appropriateness of their outputs much like supervisors direct, guide, and assess their human subordinates.

In principle, such a supervisory arrangement avoids many of the weaknesses found in human decision-making processes; the automated tools will often shield their human supervisors from the details of analysis. Very complex, time-critical decision tasks can be managed in a manner that alleviates the impact of humans' attentional limitations on decision-making. As stated earlier, the heuristics that humans use to circumvent limitations in information-processing capacity produce decisions based on partial information, which can, in turn, lead to implementing sub-optimal decisions. These heuristics are more likely to be used as the number of alternative options that must be considered increases, the number of potential consequences associated with each option increases, and the number of conflicts (or tradeoffs) among the possible consequences of the options increases. Thus, even though all the information may be available for analytical decision-making, human decision-makers don't typically work through it—especially when under time pressure. Automated tools, however, do work through the data comprehensively, managing the analytical combination of information so humans can focus attention on goals and the appropriateness of recommended or enacted options.

The knowledge and skills associated with supervision of humans, however, will not transfer directly to the supervision of automation.

New specialties and new training will be required. The new specialists will learn techniques for controlling automated tools that work with detail at a speed that exceeds human information processing capacity. The new supervisors will know how to maintain situation awareness and monitor the behavior of these tools without delving into the details of the process unless absolutely necessary. Indeed, for some of these tools, getting into the detail to maintain control will most likely be impossible. Mover and Shaker, described in Chapter 14, for example, provide technology that allows commanders to control autonomous vehicles. UV Coordinator, in Chapter 15, provides a similar capability for cooperating teams. The human supervisor, therefore, will require management strategies that use higher-level indices (task-relevant summaries and so forth) of the behavior and intentions exhibited by technologies such as these.

For other technologies, even though they perform tasks currently performed by humans, the decision-making process they use will be very different from that used by humans. Chapter 7, for example, describes technology that provides recommendations in the form of dynamically generated mission orders and proactive assignments. Its strategy for accomplishing this task, however, involves a speed and level of detail that would overwhelm a human decision-maker's attentional capacity. Part of the supervisor's task, therefore, will require mediation and interpretation of automation behavior through an interface that maps strategies that can be used by humans onto strategies used by automation and vice versa. Given that this interface is likely to be imperfect, the human supervisor will need to know when to prudently and selectively delve into details related to the automation's performance.

For these new technologies to improve C^2 system effectiveness, a great emphasis must be placed on building a cooperative work system that capitalizes on the strengths of humans and automation. Such a system must assure a smooth, effective transition between the states where the initiative belongs mainly to the human supervisor and those where the initiative is mainly that of automation. Without such a cooperative mechanism, the human-automation system can exhibit disastrous forms of failures. One example is *decompensation*, a latent system failure whose evolution exhibits a characteristic, two-phase signature [8]

In decompensation incidents, humans and automation engage in a deadly dance caused by breakdowns in coordination. In phase I, system performance gradually declines, while the automation tries to correct deviations from optimal performance. Eventually, if the human doesn't intervene appropriately, the disturbance exceeds the automation's capacity to take corrective action. At this point, phase II begins as automation hands over full control to the human operator, and the system rapidly collapses. It's at this point where the benefit gained by shielding humans from detail through automation becomes a liability. A poor interface will create a situation where there's not enough time for the human to acquire the situation awareness required for prudent action. In general, a lack of coordination between human and automation, information overload, and a poor understanding of system behavior all contribute to decompensation incidents. In addition to decompensation, other types of complex, and potentially dangerous, dynamic behaviors may occur when humans change the degree and form of supervisory intervention into automation's processes.

New Forms of Corporate Knowledge

Automation tools suggested in this book embody and act on corporate knowledge that is more sensitive to the situation than typical standard operating procedures, doctrine, and the like. These tools' sensitivity to current situations opens up the possibility that they will occasionally offer alternative, valid interpretations of the situation at hand. Thus, these technologies potentially increase the commander's mental flexibility—if used cautiously.

As commanders gain experience, they build a growing repertory of war stories in memory. When a situation reminds commanders of one of these episodes, they have a ready-built mental framework for structuring their decision-making task. This framework provides mental pegs for holding relevant information together in a meaningful way. As such, it focuses their attention on what's important, what's missing, and what's contradictory [9].

One consequence of experience is an economy of effort. The mental frameworks gained from experience highlight important relationships among the various pieces of situational data. Thus, missing and contradictory information can be targeted for additional collection and analysis activities—but only if the information is important. In addition, according to commanders' interpretation of patterns in the available data, they can make educated guesses (inferences), if necessary, to fill in missing information and resolve discrepancies.

Such mental frameworks draw attention to what needs to be done, so that a coherent understanding of the situation can emerge. Situation awareness, in turn, guides attention toward both reasonable actions and the important tradeoffs that distinguish the most appropriate actions from the alternatives—usually an action like one that's worked in a similar situation.

Commanders add structure to their decision task by consulting corporate knowledge, in the form of standard operating procedures, doctrine, and so on. Thus, commanders combine personal experience with corporate frameworks in order to facilitate decision-making processes. In a manner of speaking, the technologies described in this book also embody a type of corporate knowledge, helping structure the decision-making situation by providing a framework.

The tools exploit feedback about the rapidly changing states of enemy and friendly forces to maintain a dynamic representation of the evolving situation. In turn, this representation, like the mental framework used by experienced commanders, is a dynamic model of the players—and the important relationships among them in the perceived situation.

Commanders can consult this dynamic model to help them to decide whether a missing or contradictory piece of information is sufficiently important for additional collection activities. Commanders can also use this model to help fill in gaps in situation awareness—but without additional collection. That is, the statistical relationships in the dynamic models can be used to derive objective estimates for missing or contradictory information. Genie, described in Chapter 6, for example, uses pheromone fields to fill in such gaps about the probabilistic location of threats and targets. OpSpy, described in Chapter 5, supports inferences about enemy intent based on observable enemy movements and actions.

Although experienced commanders have always made inferences about missing or discrepant data, automation can enhance the objectivity of those inferences. The relationships used by the individual commander to make inferences are based on personal, often tacit, knowledge and experience. In contrast, the relationships encoded in automation's models may be based on the knowledge and experience of many commanders. Furthermore, the encoded relationships will include formal analyses, and thus be less susceptible to the idiosyncratic biases of one individual.

The benefit of using mental frameworks to structure decision-making tasks, however, comes with a cost. As described above, if the framework is appropriate for the situation, then, by drawing attention to what's important, that framework helps reduce the human's information-processing load. However, if the mental framework is inappropriate for the situation at hand, then it creates a mental set in which discrepant evidence is either ignored altogether or forced into an inappropriate mental model. Thus, if an inappropriate mental framework has been selected to structure a current decision task, because these frameworks tend to guide attention toward compatible information and away from incompatible information, it can therefore be very difficult to notice that the framework is indeed invalid. Therefore, the initial selection of a mental framework for interpreting a situation is of crucial importance. The implicit recognition of this importance may be why experts are more careful during initial situation assessments than novices [10].

In a similar fashion, if the technology's framework is appropriate for the situation, by drawing attention to important aspects of the situation, it, too, has the potential to reduce the human's information-processing load. Indeed, many of the preceding Chapters mention capabilities for what-if or sensitivity analyses, which can be used to guide a commander's attention to a subset of the most important situational data. Chapter 11 gives examples of such sensitivity analyses and further discusses how they can be used to help a commander. Chapters 7, 9, and 13 also describe the use of sensitivity analyses to focus a commander's attention. Several of the game-theoretic technology Chapters, particularly Chapters 12 and 13, discuss simulation-based analyses that can be used to manage attention.

Because the tools bring frameworks to the decision-making situation, they offer a mechanism for providing alternative interpretations of the situation. This mechanism, in turn, can be used to counteract the human information-processing bias of ignoring discrepant information. With the proper alerting and explanation facilities, the tools can be used to provide disconfirming (and confirming) evidence, so that commanders can continually assess the validity of their mental frameworks. These facilities, in turn, should be complimented with training regarding the search for conflicting evidence, and for heading off the possibility of creating an automation bias, which might cause commanders to trust the automation's interpretation (recommended actions, for example) too much. Training is implied, in fact, because the degree of trust, and the predisposition to believe automation, has been shown to correlate with both the reliability of the automation and the human's confidence in his or her skill. High workloads also increase the human's tendency to accept automation's recommendations or actions with only cursory assessments of their validity.

New Means for Communicating Intent

The communication of a commander's intent is perhaps the biggest challenge confronting the adoption of this book's technology concepts. For example, how will commanders express intent as objective functions, as those discussed in Chapter 12, or as probabilities of success, as in Chapter 8?

Objective functions, in this context, are equations that express in mathematical terms the tradeoffs among the various consequences that may occur as a result of a military action. Any action implemented will have multiple consequences; friendly and enemy casualties are but two obvious examples. The tradeoffs these technologies manage are based on the relative utility (that is, value, importance, or attractiveness) that commanders place on such consequences.

In a statement of a commander's intent, the value of multiple consequences may be expressed in terms of multiple objectives, such as maximize enemy casualties while minimizing friendly casualties. Most of the technologies described in this book, however, require the use of numbers to express how important maximizing enemy casualties and minimizing friendly casualties are. The relative magnitudes of these numbers will ultimately determine the recommendations or controls generated. The crux of the issue, therefore, lies in translating the commander's objectives, usually expressed in natural language, into the mathematical equations these technologies require.

Decision-making research has shown the assessment of value (that is, utility) to be very challenging, and susceptible to bias [11].

Indeed, the basic research question remains unanswered, as to whether people can actually describe the relative importance of different objectives (speed versus accuracy, for example) quantitatively. Simply changing the manner in which questions about utility are framed can systematically lead to very different valuations. In addition, people tend to have difficulties identifying all the relevant dimensions of value (important consequences), and the conflicts (tradeoffs), among the consequences. Furthermore, even if conflicts are recognized, human decision-makers often have difficulty reconciling them.

Along these lines, the technologies discussed in this book explore several different strategies by which commanders can express the utility of various consequences. Several of them (Chapter 9, for example) allow the use of relative values, thus avoiding the requirement for absolute judgments of utility. Chapter 13, for example, explores the possibility of allowing the utility of various consequences simply to be listed in priority order. Task Manager (Chapter 8) aims to avoid the issue of valuation altogether.

With regard to identifying the relevant dimensions of value, the technology concepts do offer the possibility of support—if used cautiously. The dimensions of value required by the technologies are reflections of the internal models they use. Thus, their internal models provide ready-made frameworks that identify the important dimensions of value. Commanders can use these frameworks, in turn, to determine whether a particular technology is appropriate for the decision-making task at hand. If it is, the technology's framework can be used to guide the valuation process, thus ensuring that the essential dimensions of value are considered. Furthermore, since many of the technologies can potentially use a library of models, during value assessment commanders may be able to get added support by browsing through the library. Caution must be exercised, though, so that commanders are not unduly biased by the dimensions of value deemed important by automation—as opposed to the dimensions that are truly important for the decision-making situation at hand.

New Demands for Representational Aids

Visualization, or more generally representational aiding, is an obvious requirement mentioned in virtually every one of the preceding chapters. The large amounts of data generated or used by these technologies must be fused and presented in a manner that makes it easier for commanders to interpret them. The models the technologies use to generate recommendations must somehow be made more transparent, so that commanders can gain insight into them. What's not so obvious, however, is the exact form these representational aids should take.

One thing that's seemingly not needed is another three-dimensional, animated graphic of the battlespace—in color. The F-16 pilots didn't need a 3-D display of airspeed overlaid on a full color map of geo-spatial location. Instead, what they needed was a stick-shaker, because a stick-shaker displays crucial information about the pilots' decision space. It makes perceptually salient the conceptual distinctions required for good flight control.

The basic point is that these representational aids must help commanders compare their mental models of the decision-making situation to the models of the situation used by the automation. This comparison is crucial, so that commanders can decide whether their perception of the situation, and the automation's perception, is compatible or discrepant. If they're compatible, then chances are that the recommendations will seem reasonable. If they're discrepant, questions will arise as to which model is correct, the commander's or the automation's?

In a manner of speaking, the technologies discussed give commanders rational advice deduced from sophisticated algorithms. To the untrained eye, their methods of deduction for making recommendations may appear mystical, if not outright perplexing. Even trained commanders who use this automation may find themselves asking a lot of questions, such as:

—Why did it recommend that?
—Is that recommendation valid in this case?
—Is it working properly?
—What is it doing?
—What will it do next?
—Did I accurately convey my intent to it?
—When should I use it?

Hooks, in the form of representational aids and supplemented with explanation facilities, must be placed in automated tools, so that the people using them can get their questions answered. Mechanisms for looking under the hood (for example, visualization and explanation facilities) must be provided, which support the comparison of the commander's situational knowledge with the automation's model of the situation.

SUMMARY

Perhaps the greatest challenge facing this book's technology concepts is the profound relationship between humans and automation. The tools envisioned here seem to invade the domain of life-and-death criticality, and to address aspects of military decision-making that just a decade ago were considered unsuited for automation. For such tools—decision-support aids with various degrees of autonomy—to be effective, automation must be a true partner in a distributed decision-making process. However, such a partnership is a notoriously difficult requirement, which a majority of today's systems fail to meet—even *without* the technical sophistication implied by the concepts presented here.

Even if the partnership requirement is met, the role of humans will change dramatically. Automation will help reduce the human workload at the lower level of abstraction: instead of creating a solution, the human supervises the decision-making process and evaluates alternative solutions. New jobs will not correspond directly to positions currently found in military organizations and doctrine; instead, new jobs will involve new tasks, new knowledge—and new possibilities for human and system failures.

Human mental frameworks for structuring decision-making tasks are unlikely to be consistent with frameworks used by automated tools—thereby requiring means to resolve such differences. Automation will also contain a powerful body of corporate knowledge, expressed in terms more compatible with automation's requirements than with humans'. The humans who work with these technologies will require methods for conveying their intent to the automation. The automation, in turn, must be outfitted with representational aids that organize and convey task-relevant, automation-related information to human operators. While visualization is among the best ways to conveying automation's massive amount of information to humans, many of today's visualization techniques are counterproductive. An effective representational aid may have less to do with attractive colors and maps, and more with a strong focus on the right information in the right decision space.

CHAPTER 17

PROSPECTIVE:
THE CENTRAL IDEAS

This final chapter serves as a hilltop from which the reader can not only look back over terrain traversed, but also forward to mountains not yet climbed. The goal now is to extract a few central ideas—conceptual and operational nuggets—that can serve to organize this material and direct further research, development, and deployment.

This book began with a hypothesis: control theory and its supporting technologies have developed sufficiently to be applicable to problems in C^2. Now that all the arguments are in, what conclusions can be drawn concerning this hypothesis? If it is true, control theory will have a significant, perhaps revolutionary, impact on both the organization and concepts of operation of future C^2 systems. If, however, the hypothesis fails to meet the burden of proof, then any revolution in the practice of C^2 will have to come from some other source. Otherwise, there will be only incremental improvements over current practice.

If the contents of this book are given any credence whatsoever, then that burden of proof has been met, and control theory deserves to take its rightful place alongside other techniques traditionally applied to this problem domain: operations research, artificial intelligence, planning and scheduling tools, and so on. As such, the succeeding pages briefly reconstruct this argument, reemphasizing key ideas and approaches, highlighting those most important, and summarizing the material in a useful way.

Chapter 17 is organized around seven key analogies, each capturing some important aspect of future C^2 operations concepts. Together, the analogies provide a unified and comprehensive summary of what is central to future C^2 practice. The seven analogies are:

— Football versus soccer
— Prostheses
— Psychologists
— Fortune tellers
— Bookies and brokers
— Hunting dogs
— Organisms

A final section then briefly summarizes the vision for the future.

FOOTBALL VERSUS SOCCER

The analogy between the structure of an American football game, on the one hand, and a soccer match, on the other, is not original here, but it does serve to highlight several important issues that have previously arisen. Football is a classic example of an open-loop control approach: play halts, and both sides huddle separately to consider their next moves. First, strategies for the next play are developed in a centralized manner (by the coach or quarterback). Then the two sides meet in a short, violent encounter. The outcome is assessed; the process repeats. While there is a control loop of sorts, in that the strategies developed for the next play take into account the results of the preceding play, once the ball is snapped, the possibilities for adjustment and reaction are circumscribed. Clearly, innovation at the tactical level is not a virtue in football; instead, faithful and accurate accomplishment of one's assigned role is.

This somewhat rigid, highly synchronous approach is much different from soccer's more fluid structure. In the latter, everything is constantly in motion. New opportunities arise and disappear continually; and while there are pre-existing roles and plays, roles are fluid, and plays tend to be opportunistic. In soccer, then, the control loop is constantly running: it is not centralized, but is instead distributed among the players as they continually assess risks and opportunities, taking on those roles demanded by circumstances.

Soccer, therefore, is an inherently closed-loop system in a way that football is not. Rather than conceiving the control function as separate from, and sitting above, the engagement (a view characteristic of football, and of more traditional approaches to C^2), soccer's control function is inherently embedded in the system itself. Such a mode is typical of a control theoretic approach, which models a system having a controller as an integral part. Contrary to football, a control theoretic approach does not model a system, and then attempt to add externally and artificially a control function that is in any way different or separate from the system itself.

One advantage of such a control theoretic approach is that the effect of having a controller can, itself, be modeled and then taken into account in doing optimizations. For example, Chapters 6 and 9 demonstrate how one could model the impact of the control on operational stability, and how plan-to-plan stability could be built into the control law as one factor to be modeled. Thus, agility (the ability to respond rapidly to new circumstances) could be traded off against stability (the amount of variation introduced in previous plans) as well as other tactical advantages. Only a self-aware, closed-loop controller—one that knows about its own presence in the system, and takes this knowledge into account when formulating its recommended actions—could perform such a sophisticated optimization.

Another advantage of this control theoretic view of the engagement is that this view tends to distribute control decisions downward, throughout the command hierarchy, rather than keeping it tightly centralized. Chapters 5, 8, 9, 10, 13, 14, and 15 demonstrate how hierarchical C^2 structures are used not to centralize the control decision, but rather to distribute it as far down toward the tactical level as possible. The trick, of course, is to maintain close-to-optimal behaviors in such a distributed setting. That real concern notwithstanding, evidence is strong that the advantages of a rapid response to highly dynamic local conditions may far outweigh a centralized optimization that is sluggish and requires very large amounts of highly accurate state information.

Finally, a football-like view of the world tends to lead to very high degrees of specialization: blockers, punters, passers, receivers, runners, and so on. While some degree of specialization is advantageous even in soccer (goalie skills, for example, will always be at a premium), specialization also carries the disadvantage of being brittle—that is, of being inflexible when new situations arise. While the rules of football haven't changed much over the past 50 years, the conditions of armed engagement changed and continue to change dramatically. Under such circumstances, a C^2 system that is able to optimize a few different flexible units (as opposed to many highly specialized units) appears to have some advantages for future encounters.

From a control theory point of view, the key word is to be adaptive: that is, to adjust rapidly to new circumstances, both for initialization as well as continuation as the engagement progresses. In Chapters 5 and 15, for example, depending on actual circumstances (enemy capabilities and goals, terrain, friendly constraints, and so on), there are cases in which small numbers of flexible entities could be grouped together in a virtually unlimited number of combinations. Of special interest is the ability (still in research) to generate the correct hierarchical DES control structure rapidly and automatically from observed data.

The conclusion is that future C^2 systems are likely to look much more like soccer than football. In a soccer-type embedded closed-loop environment, control theory is a natural and powerful way to model and optimize performance—a point that is repeatedly borne out by the discussions presented in this book.

PROSTHESES

The idea of medical prosthetics is to assist human beings when they are disabled. Recently, the concept has been extended to include devices that enable human beings to do certain things, or to operate in key environments, that would be impossible without the device. Examples include night-vision optics, robotic arms and hands, forklifts, and sonar, among many others. What these devices have in common is that they are best considered an extension of the human body or human nervous system—an extra feature that enables action otherwise impossible.

The preceding discussions present several examples of how control theory can enable prosthetic devices for C^2. An analogous situation is the F-16 fighter, which increased aircraft agility and maneuverability by moving the center of gravity well aft of the center of pressure. The corresponding tradeoff, however, was that a human pilot no longer possessed the requisite skills to control flight. Instead, a highly sophisticated automatic flight control system (FCS) was created to relieve the pilot of this impossible task. Here, the F-16 FCS enabled the pilot to perform routinely, predictably, and safely in regimes which otherwise would have been impossible. In this sense, then, the FCS was a kind of prosthesis; and in just this way, several of the technologies discussed earlier in this book offer a kind of C^2 prosthesis.

The idea is simple. The pace and complexity of warfare is increasing: number and type of entities, time-critical targets, very high volumes of military intelligence data, very low latencies for needed control decisions, and so on. Just as the F-16 was unflyable without its automatic FCS, C^2 of the future may be unmanageable without substantial automation. Many of the approaches discussed here—

in Chapters 7, 8, 12, 13, and 15, for example—are capable, in principle, of automating significant portions of routine C^2 functions. And just as the FCS on the F-16 freed the pilot to concentrate on higher-level issues, so automating routine aspects of the C^2 system could free commanders from unnecessary detail, enabling them to focus on matters of greater significance.

Another way in which control theory can serve as a prosthesis is its ability to model sensitivities in the battlespace. The human mind is fit to process only a fairly small number of degrees of freedom. Indeed, human beings tend to filter out many factors, only focusing on the few that experience indicates are most relevant. The computer, to the contrary, is ideally suited for combing through very high-dimensional spaces, looking for factors that are of significance now. That is, the computer does not have to rely exclusively on heuristic rules based on experience. Given a model with literally hundreds of parameters, it can exploit mathematical algorithms and computational power to examine sensitivities, thereby drawing the commander's attention to high-priority issues that might have been otherwise overlooked.

Used in this way, control theory's models and mathematics are a kind of backstop or double-check on the commander's decisions. When everything is going as expected (within the stochastic bounds of its models), the machine simply whirrs in the background. However, when boundary conditions are approached (for example, when the system appears to be approaching phase change), a figurative red light goes on, alerting the commander to the circumstance, providing appropriate analysis, and suggesting possible corrective action.

Clearly, such a system's false-alarm rate is a concern. Despite such engineering issues, however, the basic idea of using the control theoretic models as cognitive prostheses is powerful and compelling—as in the concept of operations discussions in Chapters

4 and 11. Once again, the key idea is to use sensitivity analysis to reduce the very complex, high-dimensional space of armed conflict to one low dimensional, containing only the few factors that are then most relevant.

Control theory, then, can serve as the basis for cognitive prostheses, enabling future commanders to operate routinely with levels of complexity otherwise unmanageable. By using automation to address high-volume/low-latency requirements, and by using sensitivity analysis to alert to impending critical events, control theoretic solutions can significantly increase human ability to deal with the pace and complexity of future C^2 challenges.

PSYCHOLOGISTS

Whether it goes by the name of therapy, psychoanalysis, or counseling, the goal of a psychologist is to try to understand the patient's cognitive and emotional processes. With such an understanding, then, corrective action becomes possible. In the same way, some of the work discussed above has the metaphoric potential to psychoanalyze the opposing commander. Using game-theoretic techniques, control theory can:

— Formulate models for a rational adversary's decision process
— Use such models to propose hypotheses to account for observed behavior
— Validate these hypotheses by testing future observed actions against predicted actions
— Adjust or reoptimize friendly actions to take new information into account

While this process, the discernment of enemy intent, is most thoroughly discussed in Chapter 4, several other approaches—in Chapters 11, 12, and 13, for example—are capable of supporting similar kinds of analysis, a kind of psychoanalysis of the enemy.

Given this book's findings, the systematic application of game-theoretic techniques to C^2 might be an important contribution of recent research. A main benefit of this approach is that it not only computes a control strategy for friendly assets, but it also—and simultaneously—computes an optimal control strategy for the adversary. Such a control strategy, in turn, becomes a hypothesis about the enemy's behavior that can be tested by direct observation of his battlespace actions. Perhaps even more important, such informed thought about an enemy's possible actions can be fed into simulations or other models about future behavior. Thus, instead of having to rely on scripts or heuristics—the common approach to dealing with possible enemy behavior—this method provides a rational, defensible technique for generating a credible adversary. In a sense, it is not so important whether the enemy actually decides to act in the way that has been computed as optimal for him. The real point is that control decisions can be based on models of a credible opponent's feasible actions.

Interesting theoretical problems remain with this approach, in particular the extent to which such estimates—estimates generated in this way—should be fed back into the controller. To do so would give the adversary an information path into friendly control law, which might be used to the enemy's advantage—by deception, for example. A full formulation of this problem has yet to be made, and is an important avenue for future investigation.

It is vital to emphasize again the potential of the game-theoretic techniques that have been described, as well as their likely utility in future C^2 systems. Whether in estimating adversarial intent, providing a credible adversary control policy, or simply computing a conservative worst-case strategy, the discussions presented here—especially in Chapters 4, 11, 12, 13, and 14—clearly demonstrate the power of linking game theory and optimal control theory.

FORTUNE TELLERS

A good controller needs to be a fortune teller—to have a way of predicting the future. In brief, there must be a model that is capable of:

— Considering possible courses of action (or, control signals)
— Estimating what the future might be if one or another of these actions is selected for implementation
— Using an agreed-upon metric standard
— Selecting the action that appears best

This general scheme, called Model Predictive Control (MPC), is used as the underlying technology for all the approaches discussed in Chapters 6, 7, 8, 9, 10, 11, 12, and 13.

An important distinction between these various approaches lies in the modeling approach selected. Some models are deterministic; others are stochastic (that is, inherently probabilistic—see Bookies, below). Some models have a high degree of operational fidelity; others sacrifice fidelity for mathematical rigor and/or computational tractability. In addition, models have been developed at the tactical, operational, and strategic levels of the control hierarchy. Some models require frequent, high-quality military intelligence inputs; others are more robust, tolerating errors, latencies, and missing data. Thus, in considering the proposed concept of operations in which each of these technologies has been embedded, it is critical to be aware of the modeling approach that has been selected—and of its limitations.

Further, the proposed concept of operations cannot be separated from the type and quality of the model. For example, models with relatively high operational fidelity (as in Chapters 6, 7, 9, and 10) are much more appropriate for near-term transition and/or largely unsupervised operation. By contrast, other models (as in Chapters

11, 12, and 13) have significantly great mathematical rigor and analytic underpinnings—but these strengths come at the expense of modeling fidelity. Hence, concepts of operation for these technologies must currently rely more heavily on human review for reasonableness and military acceptance. Put simply, the better the confidence in the predictive modeling capability, the greater reliance placed in unsupervised—or partially supervised—operations.

A significant conclusion is that a gap still exists between models with high operational fidelity and those susceptible to sophisticated analytic and computationally tractable algorithms. In this limited sense, then, the initial hypothesis—once again, can control theory offer unique value for major advances in C^2—has not been validated. It appears that both the research and literature are still far way from fully automated controllers, capable of the high-powered optimization techniques successfully applied in other domains. All is not lost, however. Rather, the current models and algorithms must be wrapped in concepts of operation that are cognizant of their limitations. In this broader sense, then, the hypothesis has been sustained: given an appropriate concept of operations and supporting infrastructure, MPC can serve as the basis for powerful components in the C^2 hierarchy.

Once again, control theory is key to predictive modeling because control theory explicitly observes not only the dynamics of the plant, but also the existence and effects of the control process itself. Here, battlespace behavior will be determined largely by decisions made by each side. Because control theory explicitly models this decision (or, control) process, it can more reliably estimate how the plant will evolve over time. Such close, self-referential coupling—between a controller's predictive capabilities and its optimizing algorithms—is at the heart of MPC. In effect, there can't be a good predictor that has not also modeled a controller; and there can't be a reasonable controller that does not also have a credible predictive capability. Indeed, the wide use of MPC techniques among the approaches presented here bears out this idea.

To summarize, prediction is a critical component of control, and the quality of the predictive (or, fortune telling) capability depends in large part on the quality of the associated models. Here, a tradeoff still remains between a model's operational fidelity and its mathematical rigor and computational tractability. To address this difficulty, work continues to develop suitable concepts of operation for the various technologies that are adapted to their respective modeling strengths and weaknesses. For the foreseeable future, such tradeoffs will doubtless continue to characterize applications of control theory to military operations.

BOOKIES AND BROKERS

A major theme in the previous discussion is uncertainty—how to model it, how to cope with it in military operations. Important sources of uncertainty include:

—Measurement and sensor error or latency
—Unpredictability of engagement outcomes
—Weather
—Equipment or component failures
—Changes in strategy (friendly or enemy)
—Deception

Of course, uncertainty is the enemy of prediction and, as discussed in the previous section, prediction is inextricably bound to applications of control theory.

How have the technical approaches so far reviewed dealt with uncertainty? The most common approach is to try to estimate the odds. That is, investigators attempt to estimate the likelihood of various possible sequences of events, and then use these probabilities to compute statistical metrics—expected values, for example, or standard deviations. Control decisions are then selected based on these values' ability to influence probable outcomes. There are, of course, no guarantees—only trustworthy estimates of the likelihood of possible results. In effect, the tools act like bookies, estimating odds, placing bets (or, control decisions) on the most favorable alternatives. Examples of controllers explicitly modeling uncertainty in this way include Chapters 7, 8, 9, 10, and 11.

Another approach is to treat the uncertainty as a small, bounded disturbance around an entirely deterministic main trajectory. This deterministic trajectory, it is hoped, can be described using models with well-understood mathematical structures—differential equations, finite state machines, value payoff matrices, and so on. Employing such models then permits the use of powerful analytical techniques—but at the expense of explicitly modeling the disturbances. Here, the technology is acting more like a broker than a bookie: the long-term market trend is known, while daily fluctuations are simply ignored. Examples of this approach include Chapters 5, 6, and 13.

Yet another way of dealing with uncertainty is to finesse it—by enumerating all possible eventualities. Clearly, some amount of approximation is required—that is, some aspects of the full state of the battlespace will be ignored. However, as long as it is known what to do (that is, what command to issue) in every possible case, one need not worry about the probability of being in one or another particular situation. Control, here, is not so much a matter of maximizing return, but rather of ensuring that a system does not transition into disadvantageous states. Based on ideas from DES and finite automata, two technologies, in Chapters 14 and 15, have taken such an approach.

This book's examples survey the ways in which control theory can approach questions of uncertainty while still maintaining mathematical rigor. Some of these issues are better thought of as problems in state estimation—extracting an accurate picture of the plant (or, battlespace) while dealing with missing, untimely, or inaccurate data. Others, however, and particularly those dealing with predicting future events, go to the heart of the control theoretic view of the world. Clearly, it is reassuring to see so many different approaches. Now, control theory is a mature science, with much strength and many viable approaches. Current work lends increased confidence that, among the collection of techniques, an appropriate combination can be found to deal effectively with the uncertainties inherent in military operations.

One of the advantages of a formal, control-theoretic approach is an ability to quantify the impact of military intelligence reports' quality and timeliness. That is, a control theoretic approach can compute tradeoff curves showing the relationship between top-level performance metrics and military intelligence reporting characteristics. Such tradeoff curves permit a theoretically solid engineering approach to establishing military intelligence requirements.

To summarize, the work presented in this book shows a great breadth of technical approaches for dealing with the fog of war. Based on this research presented here, and the concepts of operation in which it has been embedded, it seems clear that control science can contribute significantly to dealing with the challenge of inherent uncertainty in military operations. In the future, there will certainly be increased emphasis on techniques for state estimation to supplement the algorithms for stochastic optimization that characterize much recent research.

HUNTING DOGS

Control theory has traditionally been applied to problems in which mathematical models of behaviors and constraints (often in the form of differential equations) have been developed. One of the difficulties involved in applying control theory to military operations is that human beings, and their decisions and interactions, play a very large role. And human beings are notoriously difficult to model!

One way around this difficulty is to focus on components in the battlespace—robotics or other autonomous or semi-autonomous vehicles (air, ground, or sea), for example, which are less susceptible to the vagaries of human behavior. For such entities, the control problem is considerably eased, since behaviors are, in principle and by means of actuators and local sensors, fully describable and predictably controllable. Even in such a case, however, the problem is far from easy, especially when considering issues of man-machine interface. How are human beings to interact with, and reliably control, autonomous or semi-autonomous entities in the battlespace?

One model, or analogy, for how such interactions might take place is the hunting dog. The dog accompanies (or, travels with) the hunter, and is passive until the hunter releases it to perform a task (flush the quarry, retrieve it, and so on). The key idea is that the dog is under the direct control of the hunter until sent for a specific service. During the dog's performance, it acts largely autonomously—yet within limits imposed by training. In the same way, it is argued, being released by a human agent to perform a specific task, autonomous battlespace entities (or groups of entities) can tactically support warfighters.

Hunting dog is therefore a powerful and compelling model for how human and autonomous entities can co-exist and complement each other in a military engagement. Tracing the required supporting technologies, it is no surprise that control theory (as discussed, for example, in Chapters 14 and 15) is a key enabler. A higher-level DES (implementing the abstraction of tasks) could even be sitting above a series of task-specific controllers (implementing low-level actions and control activities to carry out tasks). The human selects the task, the high-level DES switches to the appropriate state, and the low-level controllers optimize and execute it. Such a hybrid approach, combining elements of artificial intelligence and traditional continuous control, is a very attractive architecture for creating hunting dogs with which humans can comfortably and predictably interact.

The next level of complexity concerns groups of entities cooperating to achieve a military objective. An area of research not addressed in this volume—group cooperation—is a natural extension of the previous ideas.

To summarize, an important concept of operations for the control theoretic techniques introduced here is the deployment of autonomous or semi-autonomous entities cooperating with each other and with humans to accomplish a military objective. The use of hybrid architectures, combining features from both artificial intelligence and control theory, is a very promising approach to constructing such a capability, one that will be a very active area of research and development over the next few years.

ORGANISMS

The final analogy serving as a key conceptual organizing principle is a biological one—the idea that C^2 systems and their supporting technologies may need to borrow and exploit the behavioral mechanisms of natural organisms. These ideas have been circulating for the past decade or so under such names as complexity theory and emergent behavior. The notion is essentially that a simple set of rules, followed more or less blindly by a large group of local agents, can result in complex adaptive behaviors when viewed globally. One example, among many, is the ability of insect colonies to locate food sources and construct minimal paths between the nest and the food. Those rules followed by individual insects are extremely simple, and include a substantial stochastic component. The resulting behavior, however, solves a fairly complex optimization problem, a minimal spanning tree. Could similar techniques be used successfully to control aspects of military operations?

Such solution approaches might indeed be preferable to more traditional mathematical techniques for a number of reasons. First, these methods tend to be robust in dealing with the loss of individual agents or local geographic anomalies. Second, they are also naturally adaptive: depleted food sources are forgotten, and new sources given priority, simply and in a manner that does not require extensive centralized computation or decision making. Here, implementation is inherently distributed: the information of interest is geographically dispersed—and is never collected at a single, central point. Rather, agents (here, ants) interact with the information locally, blindly following simple rules based on local information, their actions cumulatively resulting in sophisticated global behavior. With the ants, the algorithm is secure, since no single location knows more than a small piece of the total solution. For these and similar reasons, there is significant military interest in determining whether such approaches can be made operational.

In Chapter 5, a technique has been proposed based on the analogy of insect pheromones. Threats and targets in the environment (or, battlespace) are sources of synthetic pheromones. Simple place agents scattered throughout the region of interest do the local bookkeeping on pheromone strength, evaporation, and dispersion. Active agents (either real physical agents, or their algorithmic surrogates) interact with the place agents to climb pheromone gradients toward targets while avoiding threats. The result is a complex assignment of assets to targets, including optimized routing, that has the desirable properties outlined above. Extensive software simulations have demonstrated the utility of this approach against a wide variety of threat and target scenarios.

In this view of the world, the centralized commander has become significantly less important. The agents (or, sensors and weapon systems) interact directly with the local environment, as captured by the far-flung network of simple local place agents. The most radically decentralized concept of operations of all those discussed above, the organisms model has the greatest potential for revolutionizing traditional C^2 organizations and practices. In many ways, it best fits a largely autonomous force structure, with relatively large numbers of simple, inexpensive platforms. Is such a brave new world what lies ahead? Only time will tell, but the investigations presented here show convincingly that the technological underpinnings for such a system are indeed achievable.

To summarize, biological analogies based on the behaviors of natural organisms offer a powerful and robust set of new approaches to the C^2 of military operations. The example in Chapter 5, based on the analogy of insect pheromones, is only one of many such attractive research directions. Nevertheless, it clearly demonstrates how such ideas can be applied, further indicating the advantages of robustness, optimality, adaptive behavior, and security that they embody. Despite the radical characteristics of the associated concept of operations (when compared to current practice), continued research and development of such approaches is both recommended and expected.

A VISION FOR THE FUTURE

This book began with a hypothesis: control theory and its supporting technologies have developed sufficiently to be applicable to problems in C^2. The goal throughout has been to argue in favor of this hypothesis, not so much from the standpoint of technology, but from the concept of operations—the C^2 functionality—enabled by the technology. That is, the military value of the technology can only be understood based on the organization, doctrine, and infrastructure in which that technology is embedded. Thus, each of Chapters 4 through 15 contains a story about what C^2 might look like if a particular technology were brought to full fruition. Together, those stories paint a rather amazing—and unexpected—picture, capturing a vision of C^2 which is compelling, and which can serve as a beacon to guide further research and development.

Here is a short photo album of this journey—highlights that capture the most interesting and promising ideas and applications. What's in that album? Looking back, what prominences rise above the common level to draw attention and motivate further effort?

One highlight must certainly be the possibility for autonomous, or semi-autonomous, entities in the battlespace. This vision, which has arisen independent of the work presented here, can certainly benefit from the ideas developed in this book. Control theory, supplemented with artificial intelligence in a hybrid system architecture, provides a clear path for achieving ever-greater degrees of autonomy and coordinated action. Indeed, autonomous entities will be part of the battlespace of the future, and control theory offers powerful tools to support the C^2 systems such autonomous entities will require.

Another such tool is the use of game theory in optimized control. Indeed, the most compelling aspect of game theory is its ability to model, simulate, and analyze a worthy opponent who is not necessarily symmetrical. Here, decision-support tools analyze an adversary's intent by comparing predicted versus observed action. Closely coordinated groups of entities encounter both symmetric and asymmetric opposition, and ABGs construct elaborate, multi-move strategies able to respond to all possible adversary counter moves. In short, the C^2 systems of the future will increasingly rely on the power of game theory embedded in optimizing controllers.

Yet another major theme has been an ability to deal with uncertainty—in the outcome of engagements, type and arrival of new target opportunities, actions of the opponent, and timeliness and quality of military intelligence input. Under such conditions, predicting the future gives way to predicting possible futures, assessing their likelihood, and optimizing control decisions to maximize the friendly goals for the most probable outcomes. Here, technologies enable powerful decision-support tools to assess and analyze such multi-branching pathways into the future. They are able to husband scarce resources, foregoing short-term gains for the likelihood of even greater gains as the battle unfolds. And, finally, they are able to perform powerful examinations of the complex, high-dimensional battlespace, drawing the commander's attention to the small number of factors of current high importance.

These three, then—increased autonomy, game theory for modeling the active adversary, and techniques for dealing with uncertainty— stand out as the most significant accomplishments, and the most likely components, of the C^2 systems of the future. However, not all has gone smoothly. There have been significant technical obstacles, difficulties that have yet to be resolved, hurdles that must be surmounted if the vision outlined here is to become a reality. Of several previously discussed, one stands out as a great concern: lack of credibility in modeling entities and their dynamics. As such, there appears to be a fundamental tradeoff between realistic, believable models and powerful mathematical analysis and algorithms. The more realistic and validated the models, the less susceptible they are to the full analytic power of the mathematical tools of control theory. And, similarly, when the problem has been cast in a manner suitable for rigorous mathematics, the statement seems to have been at the cost of simplifications and abstractions.

One possible answer to address these modeling concerns lies in the concept of operations. Reduced fidelity in the modeling leads to concepts of operation that require human review and independent confirming evidence. It is as if the algorithm suggests hypotheses about courses of action or important aspects of the battlespace, but these hypotheses must then be reviewed and validated by independent confirming evidence. By contrast, high-fidelity models can be included in systems with a much higher degree of unsupervised autonomy. Unfortunately, these approaches must still rely on heuristics, and remain open to charges of suboptimality. Based on these results, then, the development of credible models that preserve both mathematical rigor and computational tractability has emerged as a major theme for future research.

Clearly, extraordinary possibilities lie before C^2. The addition of control theory to the arsenal of tools for supporting C^2 opens up new possibilities, new ways of conducting operations, new opportunities for increased agility and stability. Clearly, the ideas presented in this book are new and significant, embodying a vision for the future of C^2 that is both revolutionary and achievable. The future is within reach—and the community of control theorists is welcomed as partners in the important work that lies ahead.

ABOUT THE AUTHORS

PRINCIPAL EDITOR

Alexander Kott, currently a Program Manager at DARPA, was a Technical Director at BBN Technology Solutions LLC at the time when this book was written. Dr. Kott has directed a number of technology research, development, and advanced application projects in C^2, logistics, transportation, planning and scheduling, distributed multi-agent systems, and advanced human-computer interfaces. His most recent research interests include issues of human-machine decision making in highly dynamic C^2 architectures. Dr. Kott's Ph.D. from the University of Pittsburgh dealt with applications of artificial intelligence to mechanical engineering. In addition to his role as principal editor, Dr. Kott authored Chapters 4 and 5, and is a contributing author for Chapter 3.

AUTHORS

A primary author and one or more contributing authors wrote each chapter. The primary author researched each technology or topic, interviewed researchers, and authored the majority text of the chapter. The contributing authors can be researchers who developed a technology, Air Operations experts, or other subject matter experts who wrote the text for one specific segment of a chapter. The authors and contributing authors are identified here.

Jim Bortz, an Assistant Manager for the Knowledge Solutions Division at Emergent Information Technologies, Inc., retired from the USAF in 1995 after 22 years of tactical reconnaissance and fighter operations experience. Mr. Bortz has a B.S. in physics from The Pennsylvania State University, an M.S. in public administration from Troy State University, and is a graduate of the Air Command and Staff College. Mr. Bortz contributed the concept of operation for Chapter 11.

Martin J. Brown, Jr., currently a Program Manager with BBN Technology Solutions LLC, has a military and professional career that includes assignments in crisis planning and C² at all organizational levels. Mr. Brown holds a B.A. in history from Drake University, an M.A. in management from Webster University, and an M.B.A. from Webster University. In addition, Mr. Brown graduated from Squadron Officer's School, the Army Command and General Staff College, and the Air War College. Mr. Brown authored Chapter 6.

Owen Deutsch, a Task Leader and contributor for the Charles Stark Draper Laboratory, has participated in many projects involving simulation and analysis, including advanced precision weapons, GPS-guided weapons in the presence of GPS jamming and with ECCM, adaptive array antennas, missile defense interceptors, communication systems, and large-scale distributed systems involving thousands of sensor and weapon elements with distributed battle management. Dr. Deutsch received his B.S. from Columbia University, and S.M. and Ph.D. from the Massachusetts Institute of Technology. Dr. Deutsch authored portions of the HELPS vignettes, as well as other technical information in Chapter 9.

Larry Ferguson, a Subject Matter Expert for Emergent Information Technologies, Inc. and also a Pilot for the Federal Express Corporation, retired from the U.S. Air Force after a successful career as a pilot, trainer, and manager. He is an expert in tactical deception and large-force deployment and employment. Mr. Ferguson holds a B.S. in aerospace engineering from the University of Florida and an M.S. in business management from Central Michigan University. In addition, he graduated from the Pilot School at Craig

AFB and the USAF Fighter Weapons Instructor School in Florida. Mr. Ferguson supplied the concept of operation for Chapters 7, 12, and 13.

Major Sharon Heise, U.S. Air Force, a Program Manager in the Information Technology Office of the Defense Advanced Research Projects Agency (DARPA), has research interests including control of autonomous, hybrid, and distributed systems, large-scale enterprise control, and unconventional applications of systems and control theory. Major Heise received a B.S. in aeronautics and astronautics from the University of Washington, an M.S. in astronautical engineering from the Air Force Institute of Technology, and her Ph.D. in control engineering from Cambridge University, England. She is the author of the Preface and Chapter 17.

Paul Hubbard, a Control Systems Engineer at BBN Technology Solutions LLC, has current research interests including the dynamics of distributed decision making and control, and modeling and simulation of DES. Dr. Hubbard has a B.S. and M.S. in mathematics and engineering from Queen's University, Canada, and a Ph.D. in electrical engineering (systems and control) from McGill University. Dr. Hubbard authored Chapters 10, 14, and 15.

Bruce H. Krogh, a Professor in the Department of Electrical and Computer Engineering at Carnegie Mellon University, has, for the past 20 years, contributed to the theory and application of supervisory control for systems ranging from autonomous mobile robots to semiconductor manufacturing processes. His current research interests include synthesis and verification of embedded control software, distributed control strategies, Markov decision processes, and discrete event and hybrid dynamic systems. Dr. Krogh received a B.S. in mathematics and physics from Wheaton College, and an M.S. and Ph.D. in electrical engineering from the University of Illinois, Urbana. Dr. Krogh authored Chapter 7 and is a contributing author for Chapter 3.

Michael Martin, a Cognitive Systems Engineer for BBN Technology Solutions LLC, has, for the last several years, focused on cognitive task analyses and objective performance measure development, analyzing user decisions and requirements, approaches, and problem-solving strategies for the development of software systems. He holds a B.A. in psychology, a B.S. in biology, an M.S. in human factors psychology, all from the University of Arkansas, and a Ph.D. in experimental/cognitive psychology from the University of Kansas. Dr. Martin is the author of Chapters 8 and 16.

Steve Morse, a Technical Director for SPARTA, Inc., brings more than 20 years experience in defense and intelligence research and development. His current area of interest is the application of control theory to military C^2 systems. Dr. Morse has a B.A. in liberal arts from St. Johns College, an M.S. in numerical science from Johns Hopkins University, and a Ph.D. in mathematics from the University of Maryland. Dr. Morse is the author of Chapters 2 and 9 and contributing author for Chapter 17.

H. Van Dyke Parunak, ERIM's Chief Scientist, and a Scientific Fellow in ERIM's Center for Electronic Commerce (CEC), founded the CEC's AI group and directed their research; he now leads ERIM's projects in software agents, emergent behavior, and nonlinear dynamics. He holds an A.B. in physics from Princeton University, an M.S. in computer and communication sciences from the University of Michigan, and a Ph.D. in Near Eastern Languages and Civilizations from Harvard University. Dr. Parunak contributed to the technical portions of Chapter 5.

Mike Pietrucha, an EWO in USAF HQ/XOOC (Checkmate), works in civilian life as an Operations Analyst for Emergent Information Technologies, Inc. He has 11 years experience as a fighter pilot, mission commander, strike planner, and line WSO. Mr. Pietrucha was commissioned through the AFROTC program at The Pennsylvania State University, where he received a B.A. in political science, and went to Undergraduate Navigator Training at Mather

Air Force Base. He also graduated from the Electronic Warfare School, the F-4E RTU at George Air Force Base, and the Wild Weasel School. Mr. Pietrucha produced the concept of operation for Chapters 14 and 15.

Saurabh Sircar, a Senior Engineer at BBN Technology Solutions LLC, includes among his research interests theory of discrete event systems applied to system control problems and operation control problems in distributed and networked systems. Mr. Sircar has a B.S. in physics from the University of Delhi, a B.E. in electrical engineering from the Indian Institute of Science, an M.S. in computer engineering from The Pennsylvania State University, and is a Ph.D. candidate in telecommunications at the University of Pittsburgh. Mr. Sircar is the author of Chapters 11, 12, and 13.

Robert Stumpf, a Senior Airpower Theorist for Emergent Information Technologies, Inc., and a Pilot and Check Airman for the Federal Express Corporation, retired from the U.S. Navy after 22 years in flight command, and as a pilot for staff and squadron tours. He holds a B.S. in oceanography from the U.S. Naval Academy, an M.A. in national security studies from Georgetown University, and is a Graduate of the Industrial College of the Armed Forces. Mr. Stumpf supplied the concept of operation for Chapters 4, 5, and 13.

Martin van Creveld, of the Hebrew University, Jerusalem, is an internationally acknowledged expert on military history and strategy. He has authored 15 books on those subjects, including *Supplying War, Command in War, Technology and War,* and *The Transformation of War.* He has taught, or lectured, at many institutes of higher military and strategic learning in the Western world, including the National Defense University and the Marine Corps University. He has consulted widely to numerous defense establishments all over the world, including that of the U.S. (Department of Net Assessment, Maxwell Air Force Base). Professor van Creveld received a Ph.D. in international history from the London School of Economics. He is the author of Chapter 1.

RESEARCHERS

Each technology discussed in this book was developed by a group of researchers at a corporation or academic institution. The following researchers provided support and technical guidance to assist the authors with compiling the chapters.

Chapter 4: Dr. Venkatesh Saligrama, Massachusetts Institute of Technology. Dr. Jeff Shamma, University of California, Los Angeles.

Chapter 5: Dr. H. Van Dyke Parunak, ERIM.

Chapter 6: Dr. David W. Hildum and Dr. Steven F. Smith, Carnegie Mellon University. Dr. Peter A. Jarvis and Dr. Karen Myers, SRI International.

Chapter 7: Dave Logan and Dr. Jerry Wohletz, ALPHATECH, Inc.

Chapter 8: Dr. Jan Jelinek, Honeywell Technology Center. Dr. João Pedro Hespanha, University of Southern California.

Chapter 9: Dr. Milton Adams and Dr. Owen Deutsch, The Charles Stark Draper Laboratory.

Chapter 10: Dr. Rubin Johnson, ORCA.

Chapter 11: Dr. Kazufumi Ito and Dr. William M. McEneaney, North Carolina State University (NCSU). Dr. Ben G. Fitzpatrick, Tempest Technologies, LLC. Dr. Qing Zhang, University of Georgia.

Chapter 12: Dr. Hiro Mukai, Washington University, St. Louis, Missouri.

Chapter 13: Dr. Jose B. Cruz, Jr., The Ohio State University (OSU), Dr. Marwan A. Simaan, University of Pittsburgh.

Chapter 14: Yi-Liang Chen, Rockwell Science Center.

Chapter 15: Dr. Richard R. Brooks, David Friedlander, Christopher Griffin, and Dr. Shashi Phoha, The Pennsylvania State University Applied Research Laboratory (ARL).

GLOSSARY

Adaptive control—A technique in which control strategy and parameters are adjusted during the operation to match the observed behavior of the process.

Agent (Software agent)—A software component that acts largely without supervision, for example, to find data or to perform control; exhibits a degree of intelligence and cooperation.

Aggregation—Grouping lower-level objects (such as individual platforms) into a single object (such as a team). Aggregation makes control possible at a high level without generating control signals for each object in the group.

Autonomic—Platforms operating without human supervision or with limited supervision.

Battlespace—The physical area in which a battle occurs, including the relevant entities within the area, such as friendly, enemy, and environmental entities.

Blue and Red forces—Names to indicate different parties involved in a conflict. Blue is often used to indicate the friendly forces while Red is used for enemy forces. Other colors (Green, Orange, Purple) may be used if a number of parties are involved in the conflict.

Cardinality of a model—A characterization of the number of entities being modeled in the system and the number of variables used to define the models mathematically.

Centralized control—An architecture where control decisions and signals are generated at a central node, which then issues controls to remote executing entities.

Chaotic dynamics—The dynamical behavior which is not periodic and where small changes in input or system parameters lead to large changes in both qualitative and quantitative behavior of the output.

Closed-Loop (Feedback control)—A control strategy that uses feedback—observations of the controlled system's behavior—to modify the behavior of the system.

Computationally intractable—A problem that is so large or complex that it cannot be computed (solved) in a timely manner or with a reasonable amount of computational resources.

Continuous variable—Unlike a discrete variable, can take values that differ by infinitely small amounts.

Controls (Control signals)—Commands that are issued to entities, such as platforms involved in an engagement.

Control system (Controller)—A system that generates commands to modify the behavior of the controlled system.

Convergence—In a process of iteratively adjusting a solution, such as a set of controls, the case when changes in the solution from one iteration to the next become insignificantly small.

Cost function (Objective function, Payoff function)—A mathematical expression used to calculate the value of an outcome or task.

Data fusion—Taking data from many different sources and creating a combined data representation.

Deadlock—A condition where a process cannot continue because of a circular wait for resources or information.

Decomposition—Breaking down an abstraction into more concrete details, such as an objective into tasks or a task into missions.

Deterministic—Having an outcome that can be predicted because all of its causes are either known or the same as those of a previous event

Differential equations—Equations involving differentials or derivatives such as velocity and acceleration; often used to describe the dynamic behavior of entities showing changes in states with respect to time. They may be linear or nonlinear.

Discrete event system (DES)—A technique of modeling system dynamics as a set of distinct modes (states) and transitions (associated with events) between the states.

Discrete variable—A variable the values of which belong to a set of discontinuous numerical points or symbols.

Distributed control (Decentralized control)—An architecture where control decisions and signals are generated independently or in cooperation at multiple nodes, often at the executing entities such as platforms.

Dynamic system—A system that evolves over time as a result of external inputs.

Dynamic programming—A technique where a problem is defined in a succession of stages and an optimal solution for a stage is derived from the solutions from the preceding stage.

Emergent behavior—When simple behaviors executed by numerous components of a system combine into a complex, often sophisticated and adaptive, behavior of the overall system. An insect colony is an example.

Filter—A system element that receives a noisy or corrupted signal and outputs a more accurate, reliable signal. It may act as a state estimator, noise rejector, discriminator or detector.

Finite state machine (FSM)—A technique of modeling system dynamics as a finite set of distinct modes (states) and transitions between the states.

Game theory—The study of strategies for maximizing gains and minimizing losses under prescribed rules in adversarial situations. It is often applied to solutions of decision making problems in business and military.

Gradient—A measure of the change in one quantity with respect to another. Higher gradient means more drastic change.

Grid—A polygonal (typically square or hexagonal) partitioning of the battlespace.

Heuristic—An approximate rule used to help develop a solution without technical or mathematical rigor.

Hierarchical task network (HTN)—A decomposition of high-level objectives into specific tasks, often including sequencing, priority, and mission-horizon information for each task.

Hierarchical control—An architecture where higher-level controllers control the lower-level controllers, which in turn may control either yet lower-level controllers or executing entities.

Hybrid control—A control scheme that uses both discrete and continuous variables, either in the system model or the input and output variables.

Instability—The tendency of a system to exhibit large changes in the output subject to small change in inputs.

Instantiation—A concrete instance (example) of an abstraction.

Integer programming—A class of optimization problems and corresponding solutions in which the variables can only take discrete integer values.

Intelligence data—Information about the battlespace (enemy assets and troop movements, for example) that is collected for use in planning, monitoring, and executing the friendly course of action.

Linear system—A system which satisfies linear superposition, i.e. the output due to an input which is the sum of two signals is equal to the sum of the outputs for each signal independently. Methods for analysis and design of such systems are relatively well understood.

Livelock—A situation where a system is performing actions but is stuck in an unproductive cycle of steps.

Markov chain—A mathematical model of a process in which future values of a discrete variable are statistically determined by present events and dependent only on the state immediately preceding.

Metrics—Standards used to determine how well a system performed by comparing actual results to the desired results.

Model predictive control—A technique where controls are determined by optimizing the predicted performance of the controlled process, subject to constraints, over a time horizon.

Monte Carlo simulation—A simulation that employs probabilistic or stochastic variation of situation and state changes of entities. As the simulation progresses in time, the system is said to evolve stochastically. The simulation is typically run multiple times to generate an average result.

Nash equilibrium (Nash solution)—A solution to a game-theoretic formulation of a zero-sum problem. It determines the best possible actions for each side of the game assuming that the opposing side also takes the best possible actions.

Non-deterministic—See Stochastic

Nonlinear system—A system that does not satisfy linear superposition (see Linear system). These systems are notoriously difficult to analyze.

Optimal control—Control that seeks to optimize a measure of system performance. It typically considers an objective function and a set of constraints.

Order of battle (OOB)—The identification, strength, command structure, and disposition of the personnel, units, and equipment of a military force.

Package—A group of platforms that is assigned to accomplish a specific task or mission.

Probabilistic—See Stochastic

Real-time—Operating sufficiently fast to allow generated control signals to effectively direct units engaged in battle.

Reduced-order model—A model that has been simplified by reducing the number of state variables.

Robustness—The ability of a control system to maintain

performance despite uncertainty in the system parameters or the environment.

Saturation—When a system's output no longer changes in response to further changes in input beyond a certain threshold.

Sensitivity analysis—Evaluating a control system to see how changes in one or more factors affect the performance of the system.

Stability—In general, a system is stable if it always gives responses appropriate to the stimulus. Specific cases are: a bounded input always gives bounded outputs, and a small change in input always gives a correspondingly small change in output.

State—The current status of a system, usually expressed by the current values of a set of variables or the current position in a discrete set of possible positions.

State aggregation—The process of creating simpler models by combining sets of states into a single state.

State estimation—The process of determining the state of a system based on observations.

Steady state—A condition in which all state variables either remain virtually constant or recur in a cyclical pattern.

Stochastic (Non-deterministic, Probabilistic) model—A mathematical model that captures multiple possibilities in the dynamic behavior of modeled entities. In a probabilistic model, each behavior has an associated probability of occurring.

Strategic objective—A broad objective that typically applies to a large geographic region and spans a long period of time.

Suppression of enemy air defenses (SEAD)—A mission to attack enemy air defenses to allow other platforms safe passage through an airspace.

Threat laydown—The geospatial distribution of threat entities (for example, surface-to-air missile locations).

Thrashing—Unproductive cycles of behavior, for example unnecessary switching from task to task or from plan to plan, so that excessive resources are expended simply by switching and not by any productive effort.

Time aggregation—A method to simplify a model of a system by eliminating variables in the model that change state more rapidly than they can be observed.

Uninhabited vehicles (UVs) (Unmanned vehicles)—Platforms that are not controlled by a person physically residing within the platform. Such platforms receive control signals either from an on-board automated controller, a remote automated or human controller.

Zero-sum game—A game on which the objectives of the opponents are exactly diametrically opposed—the gain of one opponent means an equal amount of loss for another.

ENDNOTES

PREFACE

1. M. Athans, et al. (eds.), "Laboratory for Information and Decision Systems," *Proceedings of the MIT/ONR Workshop on C3 Systems* (MIT Press, 1977-1986).

2. Report MTR-80W00025, *Proceedings for Quantitative Assessment of Utility of Command and Control Systems*, Washington, D.C. (National Defense University Press, 1980).

Part I: Introduction

CHAPTER 1:
COMMAND IN WAR: A HISTORICAL OVERVIEW

1. M. van Creveld, *Command in War* (Harvard University, 1985), 1-17.

2. Joshua 8:18, *The Bible*.

3. J. Kromayer and G. Veith, *Heeresewesen und Kriegfuehrung der Griechen und Roemer* (Munich, 1928), 518-519.

4. M. van Creveld, *Command in War* (Harvard University, 1985), 218-226.

5. B. Owens, *Lifting the Fog of War* (New York, 2000).

6. B. Watts, Clausewitzian Friction and the Future of War (Washington, DC, 1996).

7. For a brief summary of these plans, see *Defense Daily* (April 4, 2001), 1-2.

CHAPTER 2: CONTROL IN WAR: A TUTORIAL INTRODUCTION TO CONTROL THEORY

1. For those interested in recent developments in Model Predictive Control, an excellent survey article can be found in D. Mayne, et al., "Constrained Model Predictive Control: Stability and Optimality," *Automatica* (2000), 789-814.

2. A good book in stochastic optimization is D. Bertsekas, *Dynamic Programming and Optimal Control* (Athena Scientific, 1995).

3. Game-theoretic applications to control is an advanced topic, but a good introduction can be found in J. Cruz, Jr., "Survey of Nash and Stackelberg Equilibrium Strategies in Dynamic Games," *Annals of Economic and Social Measurement* (1975), 339-344.

4. The literature on linear systems and control is immense, but one of the best texts is in T. Kailath, *Linear Systems* (Prentice-Hall, 1980).

5. Those interested in linguistic geometry should consult B. Stilman, *Linguistic Geometry, From Search to Construction* (Kluwer Academic, 2000).

6. The use of discrete event applications for control is fairly recent, and C. Cassandras and S. Lafortune, *Introduction to Discrete Event Systems* (Kluwer Academic, 1999) is a good place to start.

7. Stuart Kauffman has made complex theory and emergent behavior accessible to non-specialist audiences. His *At Home in the Universe* (Oxford, 1995) provides both introduction and motivation for this circle of ideas.

CHAPTER 3: PATHOLOGIES IN CONTROL: HOW C² SYSTEMS CAN GO WRONG

1. Human error is one of the common explanations of military failures analyzed and criticized by E. Cohen and J. Gooch, *Military Misfortunes* (Vintage Books, 1991).

2. M. Russell's discussion of personality types among military professionals in "Personality Styles of Effective Soldiers," *Military Review* (Jan-Feb 2000), 69-74, includes the types that can be characterized as predominantly high-gain or predominantly low-gain.

3. Examples of such behavior are suggested in A. Kott, P. Hubbard and M. Martin, "Autocatalytic Decision Overload in C² Systems," *Proceedings of the DARPA-JFACC Symposium on Advances in Enterprise Control* (Minneapolis, July 2000).

4. The breakthrough near Sedan on May 13, 1940, and the ensuing collapse of the French line, is discussed by Maj. C. Pfaff, "Chaos, Complexity and the Battlefield," *Military Review* (July-August 2000), 83-86.

5. M. van Creveld (1985), 203-218, describes an episode easily characterized as thrashing: The numerous unproductive changes in plans and orders of Israel's Southern Command on October 8, 1973 during the Yom Kippur War.

6. This situation is reminiscent of the difficult negotiations described in L. Meyer, "The Decision to Invade North Africa (TORCH)," *Command Decisions*, Center of Military History, (U.S. Army, 1987), 185-187.

Part II: Military Intelligence

CHAPTER 4: DISCERNING THE CODE OF ENEMY INTENT

1. This vignette follows M. van Creveld (1985), 84-96.

2. S. Gentry, V. Saligrama and E. Feron, "Dynamic Inverse Optimization," text appeared in the *Proceedings of the 2001 American Control Conference*, Arlington, VA, June 2001.

3. See [2].

4. L. Ljung, *System Identification: Theory for the User* (Prentice-Hall, 1986), provides a broad background. For applications of inverse optimization techniques in other fields, such as econometrics, see J. Rust, *Structural*

Estimation of Markov Decision Processes, and R. Engle and D. McFadden, *Handbook of Econometrics* (Elsevier Science, 1994), 3081-3142.

5.   The MACE system, produced by the Training, Exercises, and Military Operations (TEMO) Simulations Laboratory at the National Simulation Center in Fort Leavenworth, KS, is a force-on-force simulation that supports course-of-action analysis, planning, and training.

CHAPTER 5: HIGHLY DECENTRALIZED INTELLIGENCE AND C²

1.   Our choice of the name Genie for a class of locality-associated agents was inspired by *genius loci,* a Roman term for a spirit populating a particular location, and by the Arabic *jinn,* a spirit often confined to a bottle-like vessel.

2.   H. Parunak, et al., "Mechanisms and Military Applications for Synthetic Pheromones," *Proceedings of Workshop on Autonomy Oriented Computation* (2001).

3.   H. Parunak, "Go to the Ant: Engineering Principles from Natural Agent Systems," *Annals of Operations Research* (1997).

4.   S. Brueckner, "Return from the Ant: Synthetic Ecosystems for Manufacturing Control," Thesis at Humboldt University Berlin, Department of Computer Science (2000), *<www.anteaters.net/~sbrueckner/ publications/2000/thesis.pdf>.*

5.   H. Parunak and S. Brueckner, "Ant-Like Missionaries and Cannibals: Synthetic Pheromones for Distributed Motion Control, " *Proceedings of International Conference on Autonomous Agents* (2000), 467-474.

6.   We introduce the generic term Uninhabited Vehicles to refer to a variety of autonomic devices—UAV, UCAV, UGV, UUV, and so on—regardless of their medium of movement (air, ground or water) and regardless of the mission (combat, reconnaissance, and so on).

7. For a recent critical discussion of the Igloo White Operation, see P. Edwards, *The Closed World: Computers and the Politics of Discourse in Cold War America* (MIT Press, 1996).

8. H. Parunak and S. Brueckner, "Entropy and Self-Organization in Multi-Agent Systems," *Proceedings of the International Conference on Autonomous Agents* (2001).

9. J. Sauter, H. Parunak and S. Brueckner, "Tuning Synthetic Pheromones With Evolutionary Computing," *Proceedings of Workshop on Evolutionary Computation and Multi-Agent Systems* (2001).

10. J. Arquilla and D. Ronfeldt, *Swarming and the Future of Conflict* (Santa Monica, CA: RAND 2000); and S. Edwards, *Swarming on the Battlefield: Past, Present, and Future* (Santa Monica, CA: RAND 2000), <*www.rand.org/publications/DB/DB311*>.

Part III: Operations Planning

CHAPTER 6: AGILE PLANNING AND SCHEDULING

1. For additional information on the Artificial Intelligence Domain Independent Planner, see D. Wilkins, et al., "Planning and Reacting in Uncertain and Dynamic Environments", *Journal of Experimental and Theoretical AI* (1995), 197-227; D. Wilkins, "Can AI Planners Solve Practical Problems?" *Computational Intelligence* (1990), 232-246; and D. Wilkins, *Practical Planning: Extending the Classical (AI) Planning Paradigm* (Morgan Kaufmann, 1988).

2. For a detailed discussion of intensity calculations, see K. Myers, "Integrating Planning and Scheduling through Intensity Adaptation" (2000).l

3. See K. Myers (2000).

CHAPTER 7: KEEPING OPTIONS OPEN

1. For further details concerning the structure and operation of DAET.

2. For further information on approximate DP solutions, see J. Wohletz, et al., "Closed—Loop Control for Joint Air Operations," *IEEE American Control Conference* (June 2000);. Bertsekas, et al., "Rollout Algorithms for Stochastic Scheduling Problems," *Journal of Heuristics* (1999); S. Patek, et al., "Approximate Dynamic Programming for the Solution of Multi-Platform Path Planning Problems," *Proceedings IEEE Systems, Man, and Cybernetics* (1999), 1061-1066; D. Bertsekas, et al., "Adaptive Multi-Platform Scheduling in a Risky Environment," *Advances in Enterprise Control Proceedings Symposium,* sponsored by the JFACC Program, DARPA/ISO, San Diego,CA (November 1999), 121-127; D. Bertsekas, et al., "Dynamic Programming Methods for Adaptive Multi-Platform Scheduling in a Risky Environment," *Advances in Enterprise Control Proceedings Symposium,* sponsored by the JFACC Program, DARPA/ISO, Minneapolis, MN (July 2000), 121-128.

Part IV: Operational Command and Control

CHAPTER 8: QUANTITATIVE MANAGEMENT OF UNCERTAINTY

1. See D. Godbole, et al., Honeywell Laboratory's *JFACC Final Report.* Note that Honeywell Laboratories refers to Task Manager as Task Commander in its technical reports.

2. More information on the Task Manager solution can be found in J. Jelinek, "Model Predictive Control of Battle Dynamics" (May 2000).

3. See J. Jelinek, "Modeling Complex Battles: Part I—Joint and Concurrent Combat Events" (September 2000).

4. For more information on the comparison with Lanchester models, see the technical report by J. Jelinek, "Predictive Models of Battle Dynamics" (February 2000).

5. Resource Allocator goes by the name of Task Group Commander in Honeywell Laboratories' technical reports. For more information on Resource Allocator, see J. Tierno, "Distributed, Multi-Battle Model Predictive Control" (June 2000).

6. For further information, see J. Hespanha, H. Kizilocak and Y. Ateskan, "Probabilistic Map Building for Aircraft-Tracking Radars," Technical Report, University of Southern California (December 2000).

CHAPTER 9: HIERARCHICAL NEGOTIATION AND DELEGATION

1. C. Allard, *Command, Control, and the Common Defense* (Yale University Press, 1990).

2. Lt. Col. D. Meyer, "Transportation Strategy," *USMC Military Review* (Jan-Feb 2001), 31-39.

3. M. Adams, "Functional Analysis/Decomposition of Closed-Loop, Real-time Work Processes," *AGARD Lecture Series 200: Knowledge-Based Functions in Aerospace Systems,* AGARD-LS-200, November 1995.

4. W. Hall and M. Adams, "Closed-Loop, Hierarchical Control of Military Air Operations," *Proceedings DARPA-JFACC Symposium on Advances in Enterprise Control*, San Diego, CA, November 15-16, 1999.

5. M. Adams, et al., "Closed-Loop Operation of Large Scale Enterprises: Application of a Decomposition Approach," *Proceedings, 2nd DARPA-JFACC Symposium on Advances in Enterprise Control*, Minneapolis, MN, July 2000.

6. C. Barth, "Composite Mission Variable Formulation for Real-Time Mission Planning," *Draper Technical Report T-1397*, Masters of Science in Operations Research, (MIT Press, June 2001). Military professionals who have reviewed the solutions produced by the controller believe they are very good. Additional details about the model fidelity versus mathematical rigor tradeoff can be found in Section 5.1 of the Draper JFACC Final Report.

7.   For experimental examples of using relative importance as a value, see
     Sections 4.1 and 8.4 of the Draper JFACC Final Report.

8.   Various modeling error robustness experiments are described in detail in
     Section 8.4.4, "Sensitivity to Modeling Errors", in the Draper JFACC
     Final Report.

CHAPTER 10: A MARKET-BASED TECHNIQUE IN C^2

1.   For a discussion of decision making in general that touches on market-
     based frameworks, see H. A. Simon and Associates, "Decision Making and
     Problem Solving," *Report of the Research Briefing Panel on Decision Making
     and Problem Solving* (National Academy Press, 1986). For more specific
     market-based decision making applied to C^2 and other fields, see <*http://
     ai.eecs.umich.edu/people/wellman/Publications.html*>.

2.   See <*www.orca1.com/OPUSinfo.html*> for more information on OPUS.

3.   In calculating an allocation's worth, Auctioneer uses a friendly asset insurance
     cost, which is an asset's value multiplied by the route planners' calculation
     of the probability of the asset's loss. For further details, see the ORCA
     JFACC Final Report.

4.   These five values (no, some, high, extreme, and critical) represent possible
     target values (1;10;100;1,000; and 10,000, respectively) as entered in the
     SEM. For further details, see the ORCA JFACC Final Report.

5.   The iteration is based primarily on cooperative auction mechanisms, as
     discussed in R. Englebrecht-Wiggans, "Auctions and Bidding Models: A
     Survey", *Management Science, No. 2* (1980), 119-142.

6.   This target database might be similar to the current MIIDSIDB
     database.

CHAPTER 11: THE LANDSCAPE OF ROBUSTNESS AND VULNERABILITY

1. NCSU researchers have applied game-theoretic methods to the JFACC project, where they studied various aspects of the problem, such as robustness, stability, filtering and estimation. For more details of the research and the results, see W. McEneaney and K. Ito, "Stochastic Games and Inverse Lyapunov Methods in Air Operations," *Proceedings 39th IEEE CDC* (Sydney, 2000), 2568-2573; W. McEneaney, "A First Foray into Meta—Controllers for Military Command and Control Applications," submitted to *Systems and Control Letters*; W. McEneaney, "Robust Game-Theoretic Methods in Filtering and Estimation," *First Symposium on Advances in Enterprise Control* (San Diego, 1999), 1-9.

2. T. Basar and P. Bernhard, *H-Infinity—Optimal Control and Related Minimax Design Problems* (Birkhauser, 1991). This book presents the ideas of Minimax games—ideas applied in AcSenTool.

3. For a good introduction to stochastic processes, see S. Karlin and H. Taylor, *A First Course in Stochastic Processes* (Academic Press, 1975).

4. The idea of Markov Chain modeling applied to model the dynamics of units in AcSenTool is provided in J. Filar and K. Vrieze, *Competitive Markov Decision Processes* (Springer, 1997).

5. A classic text presenting the concepts and ideas of dynamic programming can be found in D. Bertsekas, *Dynamic Programming, Deterministic and Stochastic Models* (Prentice-Hall, 1987).

6. The notion of value function used in the algorithm for AcSenTool was inspired by ideas from the paper written by R. Elliott and N. Kalton, "The existence of value in differential games", *Memoirs of the American Math Society* (1972).

7. The ideas of hierarchical structures for the games in HiTeAM were introduced in the book by S. Sethi and Q. Zhang, *Hierarchical Decision Making in Stochastic Manufacturing Systems* (Birkhauser, 1994).

8. The monograph, G. Yin and Q. Zhang, *Continuous-Time Markov Chains and Applications: A Singular Perturbation Approach* (Springer-Verlag, 1998) covers the applications where Markov Chain formalisms are used to model plants.

Part V: Tactical Command and Control

CHAPTER 12: OPTIMAL CONTROL AS A ZERO-SUM GAME

1. In the early 1950s, Rufus Isaacs developed Differential Games, based on the problem of forming and solving military pursuit games, while working at RAND Corporation. A detailed technical description of this technology is detailed in Differential Games: A Mathematical Theory with Applications to Warfare and Pursuit, Control and Optimization (Wiley, 1965)

2. For examples of differential game theory, see J. Lewin, Differential Games (Springer, 1994); H. Mukai, et al., "Game-Theoretic Linear-Quadratic Method for Air Mission Control," Proceedings of the 2000 IEEE Conference on Decision and Control, Sydney, Australia (2000), 2574-2580.

3. The Kalman filtering technique is very popular, and a vast body of literature exists on the principles. The Kalman filtering field resulted from R. Kalman, "A New Approach to Linear Filtering and Prediction Problems," Transactions of the ASME-Journal of Basic Engineering (1960), 35-45. For a popular graduate textbook about Kalman filtering, see R. Brown, Introduction to Random Signal Analysis and Kalman Filtering (Wiley, 1983). For proof that the Kalman filter works, see I. Rhodes, "A Tutorial Introduction to Estimation and Filtering," IEEE Transactions on Automatic Control (1971), 688-697, and a Special Issue on Applications of Kalman Filtering, IEEE Transactions on Automatic Control (1983).

4. In a game situation, unlike a typical engineering application, the knowledge of the enemy input cannot be assumed. Particularly relevant are ideas borrowed from M. Darouach, et al., "Kalman Filtering with Unknown Inputs via Optimal Estimation of Singular Systems," International Journal of Systems Science (1995), 2015-2028.

5. The details of the process are described in H. Mukai, et al., "Chapter 6 Experiment 6: Controller with a Kalman Filter for Estimation," Interim Technical Report, JFACC Project, submitted to the JFACC Project Office in DARPA, Washington University, St. Louis (2001), 103-115.

6. A detailed discussion of this algorithm is available in H. Mukai, et al., "Sequential Linear Quadratic Method for Differential Games," to appear in the Proceedings of the 2001 American Control Conference, Washington D.C.

CHAPTER 13: WHEN OBJECTIVES DIFFER

1. For details of the Ohio State University and University of Pittsburgh joint research, see J. Cruz, Jr., et al., "Modeling and Control of Military Operations Against Adversarial Control," *Proceedings of the IEEE Conference on Decision and Control*, Sydney, Australia (2000), and J. Cruz, Jr., et al., "Game Theoretic Approach for Modeling and Control of an Extended Military Operation," to appear in *IEEE Transactions on Aerospace and Electronic Systems*.

2. For a summary of game-theoretic strategies, see T. Basar and J. Cruz, Jr., "Concepts and Methods in Multi-Person Coordination and Control," *Optimization and Control of Dynamic Operational Research Models* (S. G. Tzafestas, Editor), Amsterdam: North Holland Publishing Co. (1982), 351-394.

3. For a comprehensive textbook on dynamic game theory, see T. Basar and G. Jan Olsder, *Dynamic Noncooperative Game Theory*, Second Edition (London: Academic Press, 1995).

4. Nash, a Nobel Prize winner, wrote a classical reference for the Nash solution in J. Nash, "Noncooperative Games," *Annals of Mathematics* (1951), 286-295.

5. For a discussion of leader-follower strategies, see J. Cruz, Jr., "Leader-Follower Strategies for Multilevel Systems," *IEEE Transactions on Automatic Control* (April 1978), 244-255.

6. For a discussion on Stackelberg solution, see M. Simaan and J. Cruz, Jr., "On the Stackelberg Strategy in Nonzero Sum Games," *Journal of Optimization Theory and Applications* (May 1973), 533-555.

7. To review a paper on cooperative situations, see M. Simaan, et al., "Additional Aspects of the Stackelberg Strategy in Nonzero Sum Games," *Journal of Optimization Theory and Applications* (June 1973), 613-626.

8. The ideas for games in cooperative situations in terms of control theory are in Y. Ho, P. Luh and G. Olsder, "A Control-Theoretic View on Incentives," *Automatica* (1982), 167-180.

9. For information on multi-level systems, see T. Basar and H. Selbuz, "Closed-Loop Stackelberg Strategies with Applications in the Optimal Control of Multilevel Systems," *IEEE Transactions On Automatic Control* (April 1979), 166-178.

10. For more information on ordinal game theory, see J. Cruz, Jr. and M. Simaan, "Ordinal Games and Generalized Nash and Stackelberg Solutions," *Journal of Optimization Theory and Applications* (November 2000), 205-222.

CHAPTER 14: BATTLESPACE AS A BOARD GAME

1. B. Stilman, Linguistic Geometry: From Search to Construction (Kluwer Academic, 2000).

2. Y. Chen and F. Lin, "Modeling of Discrete Event Systems Using Finite State Machines with Parameters," *Proceedings of the 9th IEEE International Conference on Control Applications*, Anchorage, AK (September 2000), 941-946; and Y. Chen, et al., "Hierarchical Modeling and Abstraction of

Discrete Event Systems Using Finite State Machines with Parameters," submitted to *40th IEEE Conference on Decision and Control.*

3. B. Stilman (2000), 9.

4. There are several categories of abstract board games of varying complexity. The simplest are non-concurrent games where only one piece moves at a time.

5. The reachability relationship for pieces on the game board is part of item (2) in the definition of an abstract board game. See Stilman (2000).

6. More detail about LG can also be found in Stilman (2000).

7. For further information on supervisory control, see P. Ramadge and W. Wonham, "Supervisory Control of a Class of Discrete-Event Processes," *SIAM Journal of Control and Optimization* (1978), 206-230; and P. Ramadge and W. Wonham, "The Control of Discrete Event Systems," *Proceedings of the IEEE* (1989), 81-98. For more general information on automata and formal languages, see J. Hopcroft and J. Ullman, *Introduction to Automata Theory, Languages and Computation* (Addison-Wesley, 1979).

8. See Ramadge (1989) for further detail on parallel compositions; and Chen (2001) for abstraction/composition operations and how the parallel composition is adapted to the FSMwP model.

9. See J. Arquilla, and H. Fredricksen, "Graphing and Optimal Grand Strategy," *Military Operations Research* (Fall 1995), 3-19, which discusses a grand global strategy with pebbles (what are called pieces here) representing sets of forces able to win regional wars.

10. For a discussion of robustness in continuous systems, see C. DeSoer and M.Vidyasagar, *Feedback Systems: Input-Output properties* (Academic Press, 1975); J. Doyle, B. Francis and A. Tannenbaum, *Feedback Control Theory*

(Macmillan Publishing, 1992). Also see M. Papa, T. Heng-Ming and S. Shenoi, "Cell Mapping for Controller Design and Evaluation," *IEEE Control Systems Magazine* (April 1997), 52-57 for a discussion of the propagation of modeling errors in a partitions of a continuous state—space.

CHAPTER 15: TEAM COORDINATION

1. For a general discussion of coordination in artificial systems, see H. Simon, *Sciences of the Artificial* (MIT Press, 1981). For specific examples, see R. Brooks, "A Robust Layered Control System for a Mobile Robot", *IEEE Journal of Robotics and Automation* (1986), 14-23; and R. Arkin, *Behavior-Based Robotics: Intelligent Robots and Autonomous Agents* (MIT Press, 1998).

2. See the Pennsylvania State University JFACC Final Report.

3. For further information on the CORBA interface in the experimental testbed, see the Pennsylvania State University JFACC Final Report.

4. See B. Davey and H. Priestley, *Introduction to Lattices and Order* (Cambridge University Press, 1990), in particular, Tarski's fixed-point lemma.

5. See J. Hopcroft. and J. Ullman, *Introduction to Automata Theory, Languages and Computation* (Addison-Wesley, 1979). For a complete description of DES and supervisory control, see P. Ramadge and W. Wonham, "The Control of Discrete Event Systems," *Proceedings of the IEEE* (1989); and C. Cassandras and S. Lafortune, *Introduction to Discrete Event Systems* (Kluwer Academic, 1999).

6. See Ramadge and Wonham (1989).

7. The presentation demonstrates how an overall protocol of escorting damaged aircraft from the area of conflict is instituted. In brief, the event corresponding to partial damage is observed in one aircraft, which leads to the upper-level controller enabling the abort and return to base events in the first aircraft, and an escort event in the healthy aircraft. The damaged aircraft can then egress while escorted by the other.

8. See the several examples of libraries of FSMs and algorithms using the DES technology at <*www.utoronto.edu/*>, <*www.engr.uky.edu/~vigyan/lego.html*>, and <*www.control.utoronto.ca/people/profs/wonham/wonham.html*>.

9. On supervisors that can achieve specifications for any set of languages, see F. Lin, "Robust and Adaptive Supervisory Control of Discrete Event Systems, " *IEEE Transactions on Automatic Control,* December 1993. On supervisors that achieve a given specification in the presence of missed transitions, see the research from the Pennsylvania State University's JFACC final report.

Part VI: Conclusions

CHAPTER 16: RETROSPECTIVE: AUTOMATION AND HUMANS IN COMMAND

1. C. Wickens, *Engineering Psychology and Human Performance*, Second Edition (Harper Collins, 1992).

2. M. Scerbo, Theoretical Perspectives on Adaptive Automation (1996); R. Parasuraman and M. Mouloua, Automation and Human Performance: Theory and Applications, Lawrence Erlbaum Associates, Mahwah, NJ.

3. M. Fineberg, Human Performance in Military Command and Control (1996); S. Andriole, Advanced Technology for Command and Control Systems Engineering, AFCEA International Press, Fairfax, VA.

4. W. Goldstein and R. Hogarth, Judgment and Decision-Making: Some Historical Context (1997); W. Goldstein, et al., Research in Judgment and Decision Making: Currents, Connections, and Controversies (Cambridge University Press, 1997).

5. M. Endsley, *Automation and Situation Awareness* (1996); R. Parasuraman, et al., *Automation and Human Performance: Theory and Applicatio*ns, Lawrence Erlbaum Associates, Mahwah, NJ.

6. See [1].

7.   N. Sarter, D. Woods and C. Billings, *Automation Surprises* (1997); G. Salvendy, *Handbook of Human Factors and Ergonomics* (Second Edition), Wiley and Sons, New York.

8.   B. Fischhoff and S. Johnson, *The Possibility of Distributed Decision Making* (1997); Z. Shapira, *Organizational Decision Making,* Cambridge University Press, 216-237.

9.   D. Serfaty, et al., *The Decision-Making Expertise of Battle Commanders* (1997); C. Zsambok and G. Klein, *Naturalistic Decision Making*, Lawrence Erlbaum Associates, Mahwah, NJ.

10. See [9].

11. See [8].